HYDRODYNAMICS - CONCEPTS AND EXPERIMENTS

Edited by **Harry Edmar Schulz**

Hydrodynamics - Concepts and Experiments
http://dx.doi.org/10.5772/58509
Edited by Harry Edmar Schulz

Contributors

Sergey Chivilikhin, Tamara Pogosyan, Ivan Melikhov, Alexey Amosov, Anastasia Shutova, Dimitrios Patoucheas, Yiannis Savvidis, Chun-Wei Bi, Guo-Hai Dong, Yun-Peng Zhao, Harry Edmar Schulz, Juliana Nóbrega, André Luiz Andrade Simões, Henry Schulz, Rodrigo De Melo Porto, José Gilberto Dalfré Filho, Maiara Pereira Assis, Ana Inés Borri Genovez, Massimo Fosco Dino Materassi

CBS, Edition **2017**

Published by InTech

Janeza Trdine 9, 51000 Rijeka, Croatia

Publishing Process Manager Iva Simcic

Technical Editor InTech DTP team

Cover Designer InTech Design Team

Additional hard copies can be obtained from orders@intechopen.com

Hydrodynamics - Concepts and Experiments, Edited by Harry Edmar Schulz
p. cm.
ISBN 978-953-51-2034-6

INTECH

open science | open minds

Contents

Preface

The present book on "Hydrodynamics - Concepts and Experiments" contains six chapters, each of which devoted to a topic considered interesting to compose this book. Although the tools used in the different chapters are somewhat superposed, a first classification may be presented as: the first two chapters describe numerical models; the third and fourth chapters use experimental tools and results; and the fourth, fifth and sixth chapters present conceptual aspects of fluid dynamics. A more detailed description is furnished in the sequence.

The first chapter presents a proposal of numerical modelling of flow-structure interactions applied to flexible fishing nets. The authors describe their models as: porous-media fluid model, lumped-mass mechanical model, and coupled fluid-structure interaction model. The numerical results suggest the adequacy of the proposal.

The second chapter shows applications of trophic indexes, stochastic concepts and hydrodynamic models to biological processes in coastal waters, like algal bloom events. The authors explore alternative approaches to study coastal environments.

The third chapter concentrates on cavitating jets, exploring their application in erosion tests and in disinfection devices (inactivation of bacteria and decomposition of persistent compounds). The authors present experimental results and details of the equipment used in the different studies.

The fourth chapter considers a classical theme, the hydraulic jump, but presenting new points of view to calculate 1) the sequent depths (or conjugate depths), 2) the length of the roller or the jump, 3) the mean form of the two-phase turbulent surface, and 4) the mean height of the oscillations of the surface. New concepts and measured data are presented by the authors.

The fifth chapter presents the point of view of the Lagrangian formulation for fluid dynamics, stressing its usefulness to treat dissipation in flows. The author presents the concepts in a didactic way, allowing a "good transit" through the chapter.

The sixth chapter presents a study on small perturbations formed in viscous fluids, with low Reynolds numbers. The situations of half-space and finite-layer are considered, and aspects like relaxation, long waves and gravity effects, short waves and capillary effects, presence of solid bottom or a second fluid are discussed by the authors.

The editor thanks all authors for their efforts in presenting their chapters and conclusions, and hopes that these efforts will be welcomed by the professionals dealing with the different aspects of Hydrodynamics.

Harry Schulz
School of Engineering,
University of São Paulo
Brazil

Chapter 1

Development of a Coupled Fluid-Structure Model with Application to a Fishing Net in Current

Chunwei Bi, Yunpeng Zhao and Guohai Dong

Additional information is available at the end of the chapter

http://dx.doi.org/10.5772/58931

1. Introduction

Aquaculture is expanding all over the world, and the net cage is becoming prevalent in the aquaculture industry. Good knowledge of the hydrodynamic characteristics of the net cage has important significance to the design of the net cage and the welfare of the fish. An important component of a net cage is the fishing net, where the fish are kept and grown. The fishing net is completely submerged and infinitely flexible. When exposed to current, the fishing net changes its shape to reduce the hydrodynamic force acting on it, and the deformed net in turn affects the flow field around the fishing net. In this way, the flow field and the deformation of the fishing net interact and mutually influence each other. Therefore, it is very necessary to consider the fluid-structure interaction between flow and the fishing net when calculating the hydrodynamic characteristics of the net cage.

To investigate the hydrodynamic characteristics of net cages, much progress has been made by both experimental measurements and numerical simulations. Aarsnes et al. [1] conducted a series of physical model tests to measure the flow-velocity reduction, fishing-net deformation and the drag force acting on net cages. Lader and Enerhaug [2] performed experiments to investigate the forces and geometry of a net cage in uniform flow, and the flow-velocity reduction in the centre of the net cage was given as 20%. It was found that the global forces acting on a flexible fishing net were not proportional to the square of the flow velocity for all currents. Fredriksson et al. [3] conducted a numerical study on mooring tensions of a large fish farm containing 20 net cages while the flow-velocity reduction was estimated using a simplified control volume approach. DeCew et al. [4] investigated the submergence behavior of a fish cage in a single-point mooring system under currents by a numerical model and a reduction of 12.7% of the current velocity was applied to the aft portion of the net chamber. Huang et al. [5] and Xu et al. [6] analyzed the hydrodynamic behaviors of net cages in waves and a reduction factor of 0.85 was adopted in the numerical model to represent the retardation effect on the water particle speed after passing through each previous net.

Based on the above analysis, the reduction in current downstream from the fishing net has been taken into consideration in studying the hydrodynamic force and the effective volume of the net cage. However, only a reduction factor was estimated to represent the shielding effect of the fishing net for most of the numerical works. There has not been a numerical model focusing on the fluid-structure interaction between flow and a flexible net. As an extension of our previous works [7-8], a coupled fluid-structure model is developed based on a joint use of the porous-media fluid model and the lumped-mass mechanical model. Using an appropriate iterative scheme, the fluid-structure interaction between flow and fishing net can be solved.

2. Model descriptions

The coupled model includes two numerical models: the porous-media fluid model and the lumped-mass mechanical model. The porous-media fluid model can simulate the flow field around a plane net with no deformation, and the lumped-mass mechanical model can simulate the deformation of fishing net in certain current. The joint use of the above two models is presented to solve the fluid-net interaction problem in current. The steady flow field around fishing net can be obtained by using an appropriate iterative scheme as explained later in Section 2.3. This model has been introduced detailed in our previous studies [9-10].

2.1. The porous-media fluid model

The flow field around a plane net is simulated using a commercial computational fluid dynamics (CFD) software FLUENT from ANSYS Inc. The porous-media fluid model is introduced to model the plane net, and the finite volume method is used to solve the governing equations of the numerical model. Thus, a brief review of the porous-media fluid model is given here.

The governing equations of the porous-media fluid model are mainly the Navier-Stokes equations as follows:

$$\frac{\partial \rho}{\partial t} + \frac{\partial (\rho u_i)}{\partial x_i} = 0 \tag{1}$$

$$\frac{\partial (\rho u_i)}{\partial t} + \frac{\partial (\rho u_i u_j)}{\partial x_j} = -\frac{\partial P}{\partial x_i} + \rho g_i + \frac{\partial}{\partial x_j}(\mu + \mu_t)\left(\frac{\partial u_i}{\partial x_j} + \frac{\partial u_j}{\partial x_i}\right) + S_i \tag{2}$$

where t is time; μ and ρ are the viscosity and the density of the fluid, respectively; μ_t is the eddy viscosity; $P=p+(2/3)\rho k$, where p is pressure; k is the turbulent kinetic energy; u_i and u_j are

http://dx.doi.org/10.5772/58931

the average velocity components, respectively; g_i is the acceleration due to gravity; i, j=1, 2, 3 (x, y, z); and S_i is the source term for the momentum equation.

The porous-media fluid model is a hypothetical model which produces the same water-blocking effect as the fish net by setting the coefficients of the hypothetical porous media. For flow through porous media, a pressure gradient exists:

$$\nabla p = \mathrm{a}\vec{u} + \mathrm{b}\left|\vec{u}\right|\vec{u} \tag{3}$$

where a and b are constant coefficients and $\vec{u}$ is the flow velocity. This expression was proposed by Forchheimer in 1901 based on the Darcy law. For large porosity (e.g., an array of fixed cylinders), turbulence will occur and the quadratic term for the frictional force will completely dominate over the viscous term (the linear term) [11]. In this case, the linear term is only a fitting term which has no physical meaning and thus it can be neglected.

Outside the porous media, the source term for the momentum equation is given by S_i=0. Inside the porous media, S_i is calculated by the following equation:

$$S_i = -C_{ij}\frac{1}{2}\rho\left|\vec{u}\right|\vec{u} \tag{4}$$

where C_{ij} is given by

$$C_{ij} = \begin{pmatrix} C_n & 0 & 0 \\ 0 & C_t & 0 \\ 0 & 0 & C_t \end{pmatrix} \tag{5}$$

where C_{ij} represents material matrices consisting of porous media resistance coefficients, C_n is the normal resistance coefficient, and C_t is the tangential resistance coefficient.

For flow through the porous media, the drag force (F_d) and the lift force (F_l) acting on it can be expressed as follows:

$$\vec{F_d} = C_n\frac{1}{2}\rho\lambda A\left|\vec{u}\right|\vec{u} \tag{6}$$

$$\vec{F_l} = C_t\frac{1}{2}\rho\lambda A\left|\vec{u}\right|\vec{u} \tag{7}$$

where λ is the thickness of the porous media and A is the area of the porous media.

The porous coefficients in Eqs. (6) and (7) can be calculated from the drag and lift forces, while the drag and lift forces acting on a fishing net have strong relationship with the features of the net. So the connection between net features and flow field is built by the porous coefficients in the numerical model. Generally, the drag and lift forces of the plane net are obtained from laboratory experiments. These forces can also be calculated from the Morison equation:

$$F_d = \frac{1}{2}\rho C_d A u^2 \tag{8}$$

$$F_l = \frac{1}{2}\rho C_l A u^2 \tag{9}$$

where C_d and C_l are coefficients that can be calculated using empirical formulas [1, 12-14].

The porous coefficient C_n can be calculated from a curve fit between drag forces acting on a plane net and corresponding current velocities using the least squares method when the plane net is oriented normal to the flow. The other coefficient C_t can be obtained when a plane net is oriented at an attack angle (see Figure 1a). In this case, the coefficient C_t should be transformed into $C_{t\alpha}$ using Eq. (10) [15]. Next, $C_{t\alpha}$ can be calculated from a curve fit between lift forces acting on a plane net and corresponding current velocities using the least squares method.

$$C_{t\alpha} = \frac{C_n - C_t}{2}\sin(2\alpha) \tag{10}$$

where α is the attack angle that is defined to be the angle between the flow direction and the plane net in the horizontal plane; $C_{t\alpha}$ is the tangential inertial resistance coefficient for the plane net at an attack angle α.

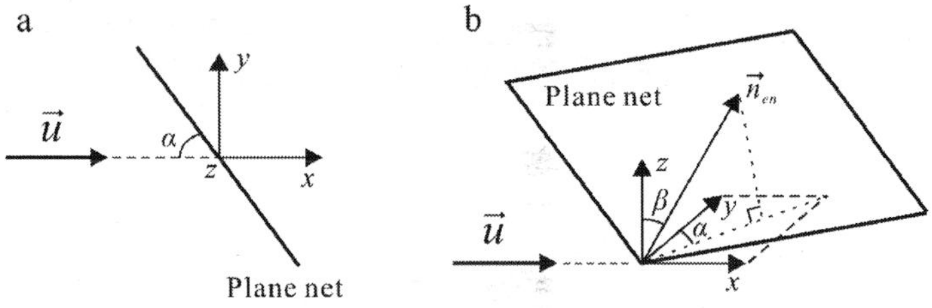

Figure 1. Definition of the angles α and β: (a) a plane net in the horizontal plane and (b) a plane net at an arbitrary position.

When a plane net is oriented at an arbitrary position (see Figure 1b), the porous coefficients should be transformed using Eq. (11):

$$\begin{pmatrix} C_n^{'} \\ C_t^{'} \\ C_t^{'} \end{pmatrix} = \begin{pmatrix} \sin^2\alpha\sin^2\beta & \cos^2\alpha & \sin^2\alpha\cos^2\beta \\ \frac{1}{2}\sin 2\alpha\sin^2\beta & -\frac{1}{2}\sin 2\alpha & \frac{1}{2}\sin 2\alpha\cos^2\beta \\ \frac{1}{2}\sin 2\beta & 0 & -\frac{1}{2}\sin 2\beta \end{pmatrix} \begin{pmatrix} C_n \\ C_t \\ C_t \end{pmatrix} \tag{11}$$

where β is the angle between the normal vector of the plane net ($\vec{n}_{en}$) and the z-direction. It should be noted that the cosines of α and β can be calculated between the corresponding two vectors.

2.2. The lumped-mass mechanical model

The lumped-mass mechanical model is introduced to simplify the fishing net, and the motion equations can be established mainly based on Newton's Second Law. Given an initial net configuration, the motion equations can be solved numerically for each lumped-mass point. Finally, the configuration of the net in current can be simulated. The calculation method for the model has been explained fully in our previous studies [16-17]. Here only a brief outline is described.

In a uniform current, the motion equations of lumped-mass point i can be expressed as follows:

$$(M_i + \Delta M_i)\vec{a} = \sum_{j=1}^{n} \vec{T}_{ij} + \vec{F}_d + \vec{B} + \vec{W} \tag{12}$$

where M_i and ΔM_i are the mass and added mass of lumped-mass point i, $\vec{a}$ is the acceleration of the point, $\vec{T}_{ij}$ is the tension force in twine ij (j is the code for the knot at another end of the bar ij), n is the number of adjacent knots of point i, $\vec{F}_d$, $\vec{W}$ and $\vec{B}$ are the drag force, gravity force and buoyancy force, respectively.

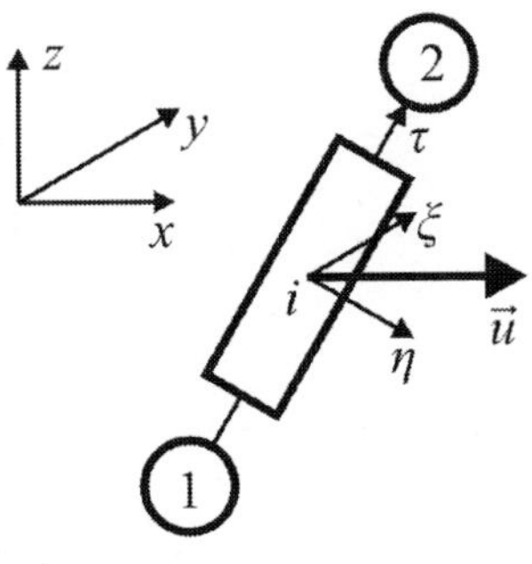

Figure 2. The local coordinate for a mesh twine.

To consider the direction of the hydrodynamic forces acting on mesh bars, local coordinates (τ, η, ξ) are defined to simplify the procedure (see Figure 2). The origin of the local coordinate is set at the center of a mesh bar, and the η-axis lies on the plane including τ-axis and flow velocity $\vec{u}$. The drag force on a lumped-mass point in the τ-direction can be obtained from the Morison equation as follows:

$$F_{D\tau} = -\frac{1}{2}\rho C_{D\tau} A_{\tau} \left| \vec{V}_{\tau} - \dot{\vec{R}}_{\tau} \right| (\vec{V}_{\tau} - \dot{\vec{R}}_{\tau}) \tag{13}$$

where $C_{D\tau}$ is the drag coefficient in the τ-direction, A_{τ} is the projected area of the twine normal to the τ-direction, $\dot{R}_{\tau}$ and $\vec{V}_{\tau}$ are the velocity components of the lumped-mass point and water particle in the τ-direction. Similar expressions can be applied to the other two components of the drag force: $F_{D\eta}$ and $F_{D\xi}$.

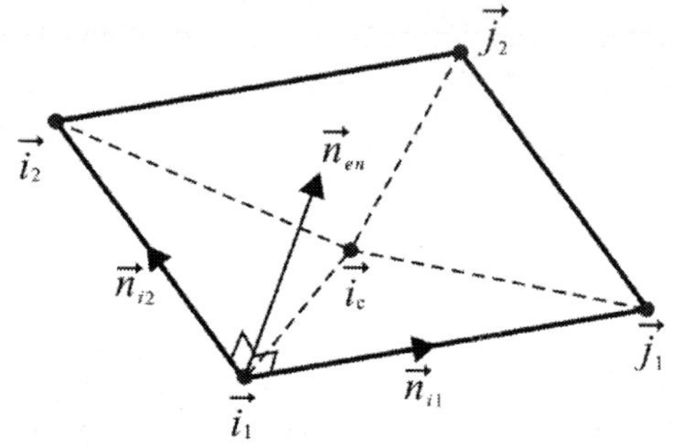

Figure 3. Schematic diagram of a plane-net element.

The mesh-grouping method [18-19] is used to reduce the computational effort of the net. Each grouped mesh is defined as a plane-net element (see Figure 3). The normal vector of the plane-net element ($\vec{n}_{en}$) is calculated as follows:

$$\vec{n}_{en} = \frac{\vec{n}_{i1} \times \vec{n}_{i2}}{\left| \vec{n}_{i1} \times \vec{n}_{i2} \right|} \tag{14}$$

where $\vec{n}_{i1}$ and $\vec{n}_{i2}$ are given by:

$$\vec{n}_{i1} = \frac{\vec{j}_1 - \vec{i}_1}{\left| \vec{j}_1 - \vec{i}_1 \right|}, \quad \vec{n}_{i2} = \frac{\vec{i}_2 - \vec{i}_1}{\left| \vec{i}_2 - \vec{i}_1 \right|} \tag{15}$$

Because of the deformation of the net, the area of a plane-net element is not constant, thus impacting the solidity of the plane-net element. Consequently, the solidity S_n must be corrected accordingly based on the varying area of the plane-net element:

$$S_n^{'} = \frac{A}{A^{'}} S_n \tag{16}$$

where S_n and A are the solidity and area of the undeformed plane-net element, and $S_n^{'}$ and $A^{'}$ are the corrected solidity and area of the plane-net element. To calculate the area of the plane-net element, the four-sided element is divided into four triangles as shown in Figure 3. The area of the element can be obtained by the sum of the four triangles.

2.3. The coupled fluid-structure interaction model

Based on the above work, the interaction between flow and fishing net can be simulated by dividing the net into many plane-net elements (see Figure 4). The main concept of the numerical approach is to combine the porous-media fluid model and the lumped-mass mechanical model to simulate the interaction between flow and flexible nets. Taking a single flexible net as example, a calculating flow chart for the numerical approach is shown in Figure 5. A more detailed calculation procedure is given as follows.

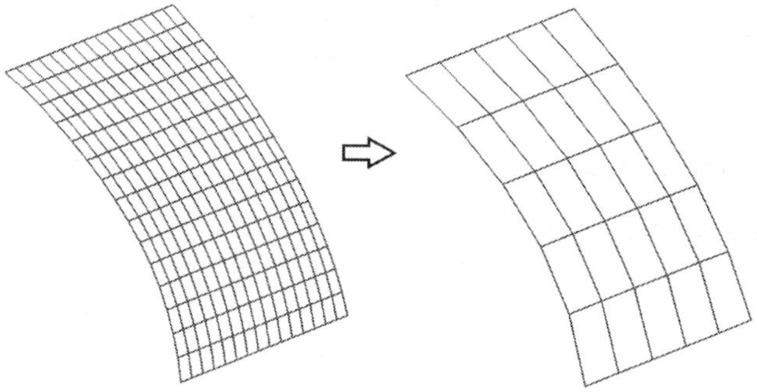

Figure 4. Schematic diagram of the net model.

The calculation procedure includes four steps. Step 1 is to simulate the flow field around a vertical plane net without considering the net deformation using the porous-media fluid model. Step 2 is to calculate the drag force and the configuration of a flexible net in current using the lumped-mass mechanical model. The given initial flow velocity is the average value exported from the upstream surface of the plane net in last step. In Step 3, according to the configuration of the net, the orientation and the porous coefficients of each plane-net element can be obtained, thus obtaining the flow field around the fishing net. In Step 4, the drag force

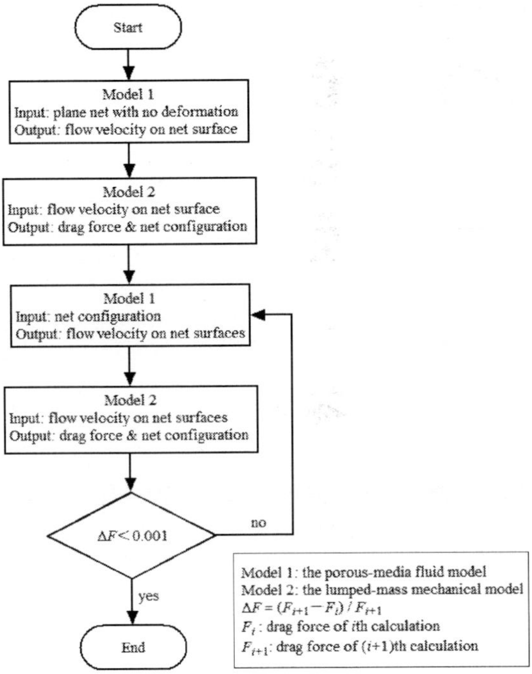

Figure 5. Flow chart of the coupled fluid-structure interaction model.

and the configuration of the net are calculated again. The flow velocity acting on each lumped-mass point is the average value exported from the upstream surface of the corresponding plane-net element in last step.

The terminal criterion of the calculation procedure is defined as $\Delta F = (F_{i+1} - F_i) / F_{i+1}$, where F_i and F_{i+1} are the drag forces of two adjacent calculations. If $\Delta F < 0.001$, it was determined that the force acting on the fishing net is stable and the configuration of the net is the equilibrium position considering the fluid-structure interaction. Thus the calculation procedure is completed. Otherwise, continue with the next iteration of the procedure and repeat Steps 3 and 4 until the criterion meets the precision requirement. Finally, the fluid-structure interaction problem can be solved and the steady flow field around the fishing net can be obtained.

3. Discussions of model parameters

Due to uncertainties in numerical simulation, convergence studies were performed to analyze the influence of different model parameters, such as thickness of the porous media, grid refinement, time-step size, number of net elements, number of iterations, porous coefficients, etc. When varying one parameter, all other parameters were kept at their initial values as shown in Table 1.

Parameter	Value
Thickness of porous media	20 mm
Grid refinement	Grid 2 [a]
Time-step size	0.04 s
Number of net elements	5×5
Number of iterations	2
Porous coefficients	C_n=35.3 m^{-1}
	C_t=10.0 m^{-1}

Table 1. Initial values of the numerical model. [a] Main properties of the medium grid are shown in Table 2.

Grid	Min. cell size (mm)	Max. cell size (mm)	Total cells
Grid 1	5	50	413,697
Grid 2	10	50	117,294
Grid 3	20	50	49,807
Grid 4	10	60	34,492
Grid 5	10	70	21,124

Table 2. Main properties of the meshes used in the grid-refinement study.

The modeled net was a 0.3 m×0.3 m knotless polyethylene (PE) net with 15 meshes in width and 15 meshes in height. The twine diameter was 2.6 mm, and the mesh bar length was 20 mm. Mounted as square meshes, the net solidity ratio was 0.243.

3.1. Thickness of the porous media

A study on the thickness of the porous media was performed for a plane net at an attack angle α=90° and incoming velocity u_0=0.226 m/s. The corresponding Reynolds number is 5.88×10^2 calculated according to the twine diameter of the net. The dimensions of the porous media are the same as that of the fishing net in both width and height, and the porous media has a certain thickness. In order to investigate the influence of the thickness on the numerical results, the porous-media fluid model with different thicknesses, λ=5, 10, 20, 30, 40 and 50 mm, were performed.

The thickness of the porous media can affect the distribution of the flow velocity around the fishing net to some extent (see Figure 6). However, the numerical results of the flow velocity are almost the same for the thickness of 5, 10 and 20 mm. It was determined that the thickness of the porous media does not affect the numerical results obviously until the thickness is greater than 20 mm. Therefore, a thickness of 20 mm is adopted hereafter for all the computations considering the total number of cells and computational accuracy. An increase in the thickness of the porous media can effectively reduce the number of cells, thus reducing the computational effort greatly. However, it is suggested that the maximum thickness of the porous media should not be greater than 10% of width or height of the plane net.

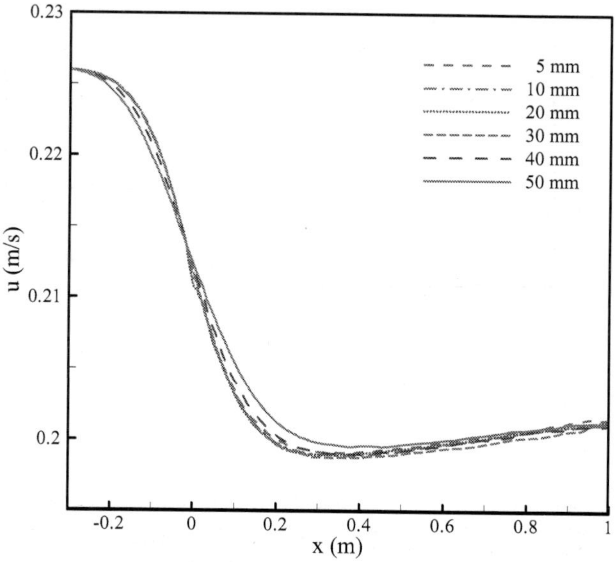

Figure 6. Numerical results of flow velocity *u* along the centerline of a plane net at different thickness of the porous media.

3.2. Grid refinement

The influence of grid on the numerical results was investigated by repeating the computations with different levels of refinement. Five grids as shown in Table 2 were taken into account. The minimum cell size was adopted inside the porous-media region. A growth rate of 1.2 was set around the net model, and the maximum cell size in the computational domain was set to control the distribution of the grid.

Numerical results of flow velocity obtained with different grids were compared (see Figure 7). It is found that numerical results are free of grid influence for Grid 1 and Grid 2. But for the coarser grid models, the numerical results tend to divergence, especially downstream from the fishing net. Therefore, Grid 2 is adopted hereafter for all the computations.

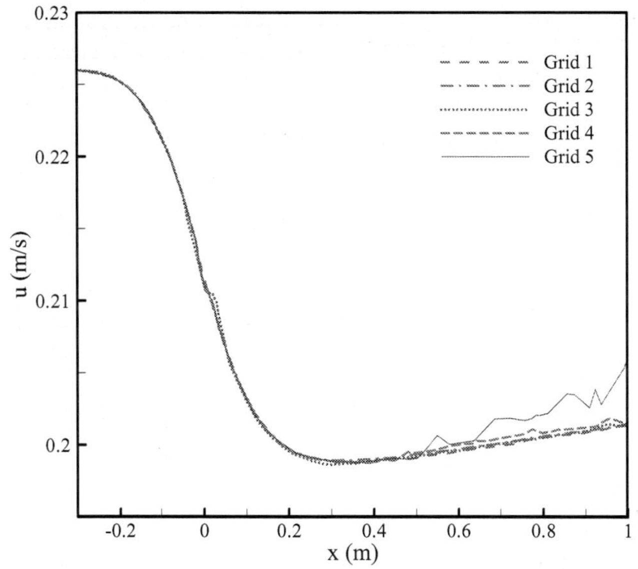

Figure 7. Numerical results of flow velocity *u* along the centerline of a plane net at different grids.

3.3. Time-step size

In order to investigate the influence of the time-step size, the computations were repeated with different time-step sizes: 0.01, 0.02, 0.04, 0.06, 0.08 and 0.10 s. The flow velocity both upstream and downstream from the plane net shows a convergence result for the time-step size smaller than 0.04 s (see Figure 8). Therefore, a time-step size of 0.04 s is adopted for computational effort saving. The computational effort decreases with increasing time-step size. However, it is suggested that the time-step size should not exceed the ratio between the minimum value of the cell size and the incoming velocity. Because the larger the time-step size is, the greater the residual error of each step will be, thus causing a divergence result without meeting the convergence criteria.

3.4. Number of plane-net elements

In the numerical model, each mesh is defined as a plane-net element. In order to reduce the computational effort, the mesh-grouping method was used to reduce the number of plane-net elements. The original model of the fishing net was a 15×15 model with 15 meshes in width and 15 meshes in height. Two grouped models were used to simplify the original model. For the original model, nine meshes were grouped into one mesh, thus obtaining a 5×5 model with a total of 25 elements. A 3×3 model with 9 elements can be obtained by grouping 25 meshes into one mesh.

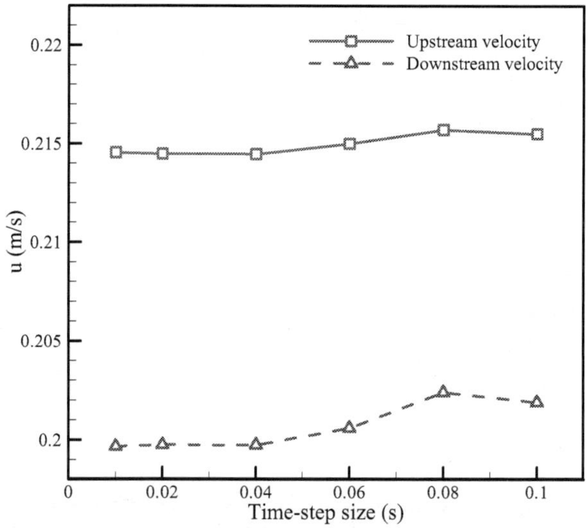

Figure 8. Numerical results of flow velocity *u* calculated with different time-step sizes. Upstream velocity is the average value exported from the upstream surface of the plane net. Downstream velocity is the flow velocity 0.6 m downstream from the center of the plane net, approximately twice the width of the plane net.

The calculated net configuration of the 5×5 model is almost coincided with that of the original 15×15 model (see Figure 9). The result of the 3×3 model agrees well with that of the original model except for a slight discrepancy at the bottom of the net. There is no significant difference between the results of flow velocity obtained with the grouped models and the original model (see Figure 10). Therefore, the mesh-grouping method can be used to reduce the number of plane-net elements without affecting the calculation accuracy. Because of better agreement on net configuration, the 5×5 model is chosen for further numerical simulations.

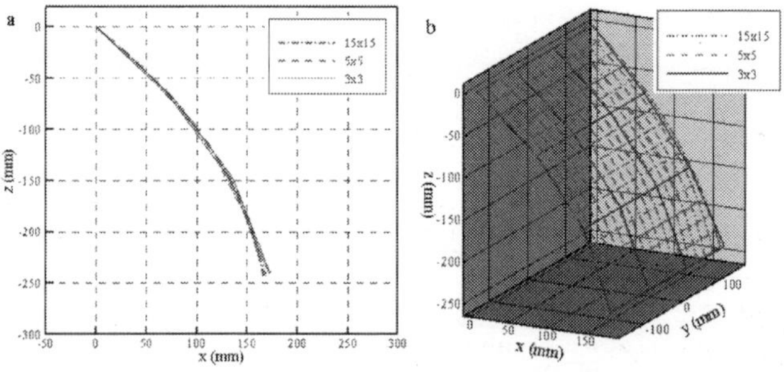

Figure 9. Comparison of the net configurations with different models for incoming velocity of 0.226 m/s: (a) side view and (b) 3D view.

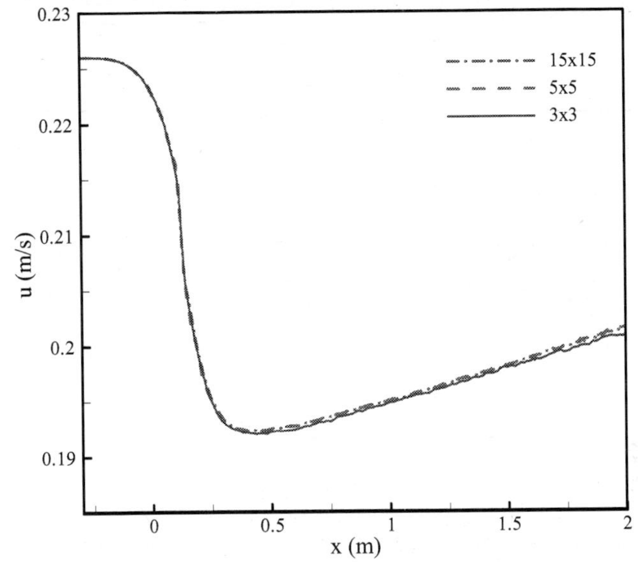

Figure 10. Comparisons of the flow-velocity component *u* along a line through the net in the *x*-direction obtained with different models for incoming velocity of 0.226 m/s. The straight line is y=0 on the horizontal plane z=–0.15 m.

3.5. Number of iterations

The interaction between flow and fishing net can be solved after several iterations. Taking a single net as example, the calculation precision reached the terminal criterion, ΔF< 0.001, within two iterations, and the calculation procedure terminated for incoming velocity of 0.226 m/s. The drag-force difference between the 1st iteration and the 2nd iteration was approximately 0.7%. One more iteration was also calculated, and the result tended to converge.

Comparing the net configurations after different iterations, there is a little discrepancy between the 1st iteration and the 2nd iteration. However, the calculated net configuration of the 3rd iteration is almost coincided with that of the 2nd iteration (see Figure 11). The flow velocity upstream of the flexible net is greater than that of the plane net with no deformation, while the flow velocity downstream from the flexible net is smaller (see Figure 12). It was determined that the deformed net has more obvious shielding effect than the plane net with no deformation. There exists rather obvious discrepancy of the flow-velocity distribution between the 1st iteration and the 2nd iteration (see Figure 12), and the maximum discrepancy is approximately 6.3%. The result of the 2nd iteration is almost the same as that of the 3rd iteration, and the maximum discrepancy is approximately 0.17%. Therefore, two iterations is adopted to model a single net for computational effort saving.

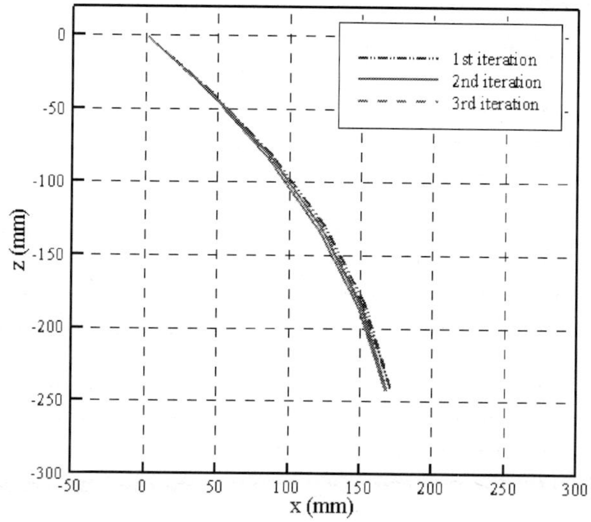

Figure 11. Comparison of the net configurations after different iterations for incoming velocity of 0.226 m/s.

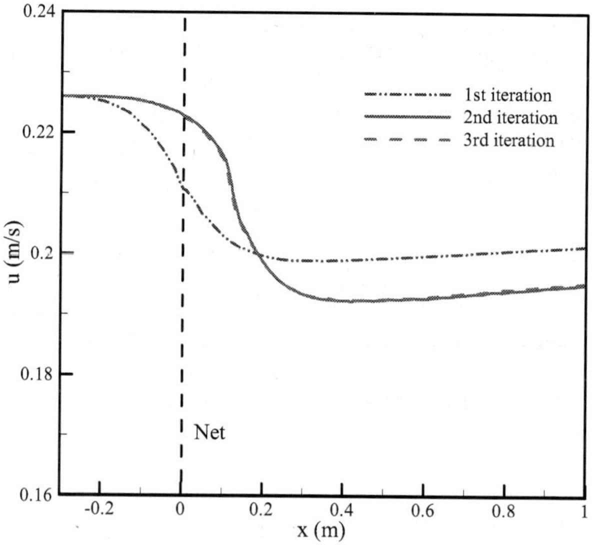

Figure 12. Comparisons of the flow-velocity component u along a line through the net in the x-direction after different iterations for incoming velocity of 0.226 m/s. The straight line is y=0 on the horizontal plane z=–0.15 m.

3.6. Porous coefficients

A set of porous coefficients, C_n=35.3 m^{-1} and C_t=10.0 m^{-1}, can be obtained from the experimental data (see Tables 3 and 4). The corresponding laboratory experiment is presented in our previous study [9]. Porous coefficients are key to the numerical results. Before the numerical simulations, a sensitivity study on the flow field with respect to the porous coefficients was carried out with a plane net at different attack angles.

u (m/s)	0.058	0.113	0.170	0.226
F_d (N)	0.082	0.395	0.968	1.678

Table 3. Drag forces at different incoming velocities in the laboratory experiments for attack angle of 90°.

u (m/s)	0.058	0.113	0.170	0.226
F_l (N)	0.032	0.131	0.321	0.534

Table 4. Lift forces at different incoming velocities in the laboratory experiments for attack angle of 45°.

When C_n is kept at a constant, the coefficient C_t increases with an increment of 2 from 6 to 12. As shown in Figure 13a, varying C_t has no effect on the flow field around the plane net for attack angle of 90°. However, when the plane net is positioned at a smaller attack angle; the flow velocity decreases with increasing C_t. As the attack angle decreases, the influence of variation of C_t to the flow velocity increases. When C_t is a constant, the flow velocity decreases linearly with increasing C_n (see Figure 13b). In addition, the variation tendencies are almost the same for different attack angles. With a same increment of 6, the coefficients C_t and C_n have different influence on the flow field around the plane net. The maximum discrepancies in flow velocity are 4.3% for the variation of C_t at an attack angle of 30° and 2.1% for the variation of C_n at an attack angle of 90° respectively.

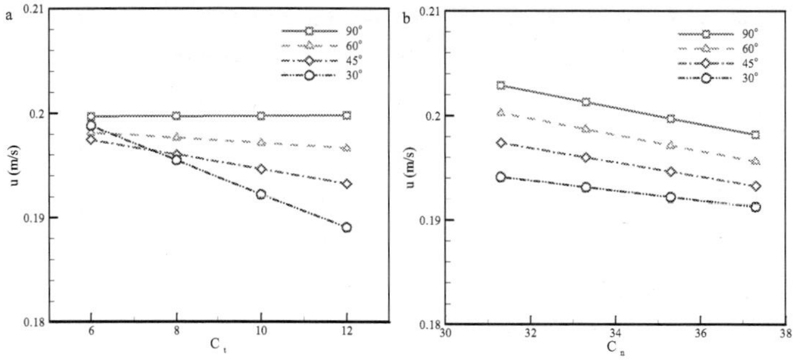

Figure 13. Comparisons of the flow velocity 0.6 m downstream from the center of the fishing net calculated with different sets of porous coefficients: (a) varying C_t for C_n=35.3 m^{-1} and (b) varying C_n for C_t=10.0 m^{-1}.

3.7. Brief conclusion

In summary, numerical results depends on the different model parameters to some extent. Appropriate values of the thickness of the porous media, the grid refinement and the time-step size are beneficial to reducing the computational effort and keeping high calculation accuracy. The mesh-grouping method can be used to reduce the number of plane-net elements and the computational effort. Two iterations are sufficient to achieve a converged solution of a flexible fishing net. The porous coefficients have a significant effect on the flow velocity around the fishing net and the relationship between each coefficient and the flow velocity is approximately linear. In addition, the influence of the coefficient C_t to the flow velocity is different with different attack angles. Based on the above, the initial values (see Table 1) are the optimal parameters for the numerical model.

4. The advantage of the coupled model

In present numerical model, the fluid-structure interaction (FSI) between flow and fishing net is taken into consideration. Because of the shielding effect of the fishing net, there exist the flow-velocity-reduction regions upstream of and downstream from the fishing net (see Figure 14). According to the numerical simulation, the reduction regions become smaller when considering FSI due to the decrease in height of the fishing net. However, the flow-velocity reduction is larger than that without considering FSI downstream from the fishing net. In the numerical results, the streamlines present the flow direction (see Figure 14). It was determined that the diversion of flow direction around the fishing net is relatively small but more obvious than that without considering FSI.

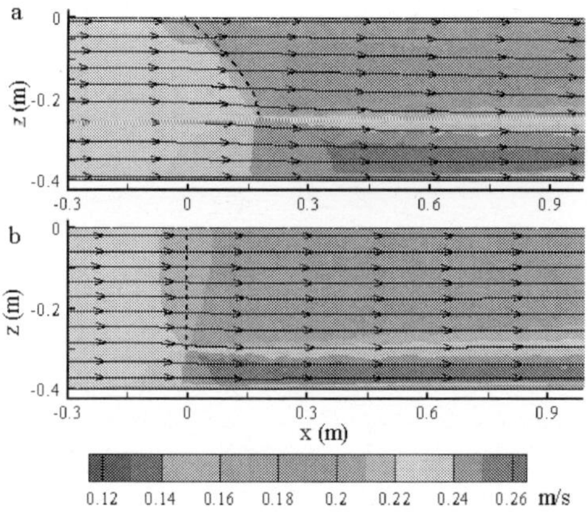

Figure 14. Flow-velocity distribution on the vertical plane y=0 around the net calculated by numerical simulation for incoming velocity of 0.226 m/s: (a) considering FSI between flow and fishing net and (b) without considering FSI.

Comparing the flow-velocity distribution along the x-direction, there exists rather obvious discrepancy between the fishing net with and without considering FSI (see Figure 15). The flow velocity upstream of the fishing net is greater than that without considering FSI, while the flow velocity downstream is smaller. The maximum flow-velocity reductions downstream are 15.0% for the FSI model and 11.9% for the model without considering FSI, respectively. It was found that the deformed net has more obvious shielding effect than the net with no deformation. This is due to the deformation of the flexible net. When exposed to current, the flexible net changes its shape and deviates from the original position. If the net is evenly divided into several plane-net elements, the angle between the incoming current and each net element (attack angle) decreases. Thus there is a relative increase in net solidity along the flow direction, and resulting in a more obvious blockage effect.

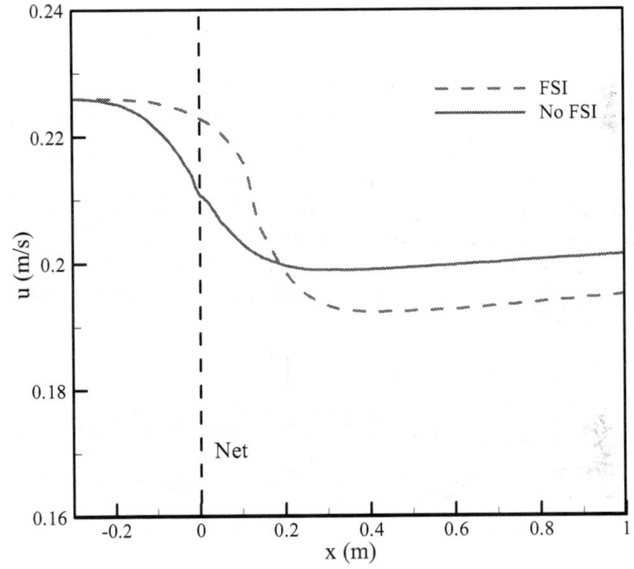

Figure 15. Comparisons of the flow-velocity component u along a line through the net in the x-direction with and without considering FSI for incoming velocity of 0.226 m/s. The straight line is y=0 on the horizontal plane z=−0.15 m.

As shown in Figure 16, the deformation of the fishing net is less serious than that without considering FSI. This is because that the net configuration is calculated based on the actual velocity acting on the flexible net instead of the incoming velocity in present numerical model. The actual velocity is slightly smaller than the incoming velocity due to the blockage effect of the net. As a result, the drag force is a bit smaller at approximately 7-8% than that without considering FSI for different incoming velocities (see Figure 17).

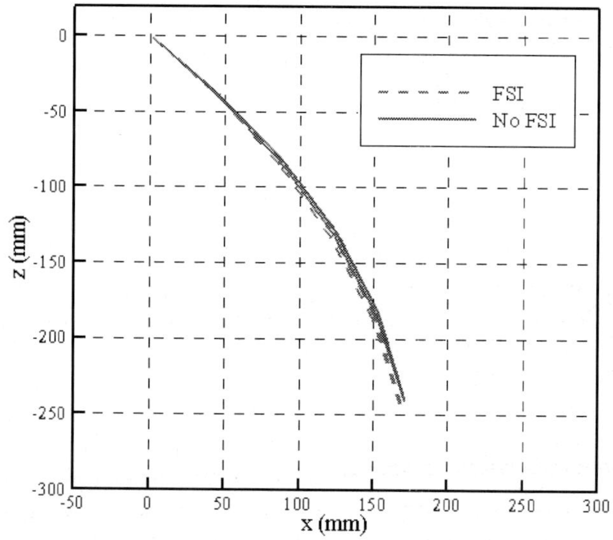

Figure 16. Comparison of the net configurations with and without considering FSI for incoming velocity of 0.226 m/s.

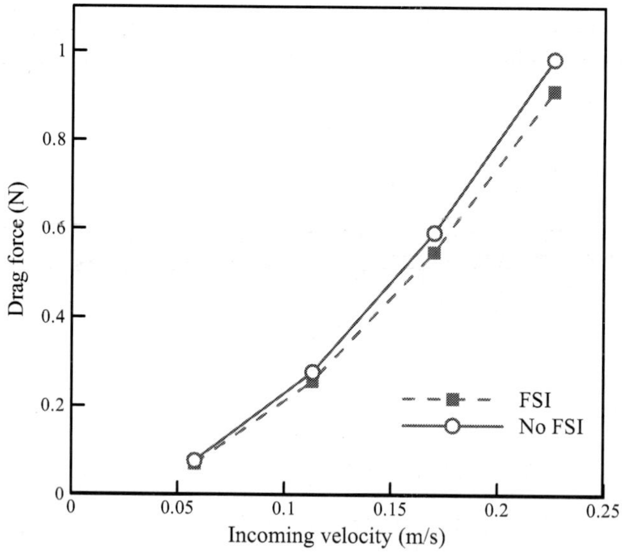

Figure 17. Comparisons of the drag force with and without considering FSI at different current velocities.

According to the analyses above, there are obvious difference between the flow field, the hydrodynamic force and the deformation of the fishing net with and without considering FSI. For multiple nets, the influence of FSI on the hydrodynamic characteristics of the fishing net will become more obvious. Therefore, it is necessary to take FSI into consideration in studying the interaction between flow and fishing net.

5. Conclusions

A numerical model is proposed to simulate the interaction between flow and fishing net based on the joint use of the porous-media fluid model and the lumped-mass mechanical model. Convergence studies are performed to analyze the influence of different model parameters on the numerical results and a set of optimal parameters are obtained for the numerical model. The numerical results of net deformation, drag force and flow velocity around the fishing net agree well with the experimental data. It is indicated that present numerical model can simulate the interaction between flow and fishing net accurately.

Compared with the numerical results without considering FSI, present numerical model is considered to present more accurate results: (i) the flow-velocity reduction is larger downstream from the fishing net, (ii) the diversion of flow direction around the fishing net is more obvious, (iii) the deformation of the fishing net is less serious and (iv) the drag force is a bit smaller at approximately 7-8% for different incoming velocities.

This study forms a foundation for studying the flow field and hydrodynamic characteristics of net cages and offers some valuable information for the aquaculture industry.

Acknowledgements

This work was financially supported by the National Natural Science Foundation (NSFC) Project Nos. 51239002, 51221961 and 51109022 and the Fundamental Research Funds for the Central Universities (DUT13LK55).

Author details

Chunwei Bi, Yunpeng Zhao* and Guohai Dong

*Address all correspondence to: Zhaoyp18@hotmail.com

State Key Laboratory of Coastal and Offshore Engineering, Dalian University of Technology, Dalian, China

References

[1] Aarsnes JV, Rudi H, Løland G. Current forces on cage, net deflection. In: Engineering for Offshore Fish Farming, Oct 17-18, Thomas Telford, London, 1990. pp. 137–152.

[2] Lader PF, Enerhaug B. Experimental investigation of forces and geometry of a net cage in uniform flow. IEEE Journal of Oceanic Engineering 2005; 30 (1): 79–84.

[3] Fredriksson DW, DeCew JC, Tsukrov I, Swift MR, Irish JD. Development of large fish farm numerical modeling techniques with in situ mooring tension comparisons. Aquacultural Engineering 2007; 36 (2): 137–148.

[4] DeCew J, Tsukrov I, Risso A, Swift MR, Celikkol B. Modeling of dynamic behavior of a single-point moored submersible fish cage under currents. Aquacultural Engineering 2010; 43 (2): 38–45.

[5] Huang CC, Tang HJ, Pan JY. Numerical modeling of a single-point mooring cage with a frontal rigid frame. IEEE Journal of Oceanic Engineering 2009; 34 (2): 113–122.

[6] Xu TJ, Zhao YP, Dong GH, Li YC, Gui FK. Analysis of hydrodynamic behaviors of multiple net cages in combined wave-current flow. Journal of Fluids and Structures 2013; 39: 222–236.

[7] Zhao YP, Bi CW, Dong GH, Gui FK, Cui Y, Guan CT, Xu TJ. Numerical simulation of the flow around fishing plane nets using the porous-media model. Ocean Engineering 2013; 62: 25–37.

[8] Zhao YP, Bi CW, Dong GH, Gui FK, Cui Y, Xu TJ. Numerical simulation of the flow field inside and around gravity cages. Aquacultural Engineering 2013; 52: 1–13.

[9] Bi CW, Zhao YP, Dong GH, Xu TJ, Gui FK. Numerical simulation of the interaction between flow and flexible nets. Journal of Fluids and Structures 2014; 45:180–201.

[10] Bi CW, Zhao YP, Dong GH, Zheng YN, Gui FK. A numerical analysis on the hydrodynamic characteristics of net cages using coupled fluid–structure interaction model. Aquacultural Engineering 2014; 19: 1–12.

[11] Burcharth HF, Andersen OH. On the one-dimensional steady and unsteady porous flow equations. Coastal Engineering 1995; 24 (3–4): 233–257.

[12] Balash C, Colbourne B, Bose N, Raman-Nair W. Aquaculture net drag force and added mass. Aquacultural Engineering 2009; 41 (1): 14–21.

[13] Tsukrov I, Drach A, DeCew J, Swift MR, Celikkol B. Characterization of geometry and normal drag coefficients of copper nets. Ocean Engineering 2011; 38 (17-18): 1979–1988.

[14] Zhan JM, Jia XP, Li YS, Sun MG, Guo GX, Hu YZ. Analytical and experimental investigation of drag on nets of fish cages, Aquacultural Engineering 2006; 35 (1): 91–101.

[15] Bear J. Dynamics of fluids in porous media. America Elsevier Publishing Company, Inc., New York, USA, 1972.

[16] Li YC, Zhao YP, Gui FK, Teng B. Numerical simulation of the hydrodynamic behaviour of submerged plane nets in current. Ocean Engineering 2006; 33(17-18): 2352–2368.

[17] Zhao YP, Li YC, Gui FK, Dong GH. Numerical simulation of the effects of weight system on the hydrodynamic behavior of 3-D net of gravity cage in current. Journal of Hydrodynamics, Series B 2007; 19 (4): 442–452.

[18] Tsukrov I, Eroshkin O, Fredriksson DW, Swift MR, Celikkol B. Finite element modeling of net panels using a consistent net element. Ocean Engineering 2003; 30 (2): 251–270.

[19] Zhao YP, Li YC, Dong GH, Gui FK, Teng B. Numerical simulation of the effects of structure size ratio and mesh type on three-dimensional deformation of the fishing-net gravity cage in current. Aquacultural Engineering 2007; 36 (3): 285–301.

Chapter 2

Modelling of Hydrobiological Processes in Coastal Waters

Dimitrios P. Patoucheas and Yiannis G. Savvidis

Additional information is available at the end of the chapter

http://dx.doi.org/10.5772/59498

1. Introduction

It is well known that the dispersion of organic and inorganic matter in a coastal basin is closely related to the circulation of the seawater masses. The more detailed, accurate and reliable the knowledge of the hydrodynamic circulation is, the more successful the initial tracing of matter distribution in a coastal basin is expected to be.

In general, marine hydrodynamics is related to many different physical, chemical, geological and biological processes, which often constitute important environmental problems that need special care and detailed investigation. Among the most frequent phenomena of the aforementioned environmental issues, observed mainly in coastal basins, are the algal blooms, which are strongly related to optical, aesthetic and other important disturbances, such as eutrophication processes. The appearance of such phenomena is becoming more and more important, because algal blooms may contain potential toxic populations, called harmful algal blooms (HAB). As such episodes may correlated with eddies development [1], their study may obviously furnish a potentially effective prediction tool. The next important step in this rather preliminary process of tracing the potential locations of harmful algal blooms is the development and application of a transport model, which computes the matter transfer in space and time. A further step is the integration of a biological model into a transport model. It is expected that mathematical models that include physicochemical and biological processes will be able to describe the marine hydrodynamics and the relevant matter transfer issues, such as episodes of the spread of phytoplankton cells, contributing to the better understanding and more effective investigation and management of similar phenomena. Such models have been applied and discussed in literature [2-6].

The models to be used depend on the available data, the specific scientific questions to be answered and the particular characteristics of the ecosystem. The mathematical models used for hydrobiological processes in coastal waters can be focused on diagnostic matters, like the trophic state and the water quality of a specific coastal zone, or on prognostic matters, like the dispersion of HAB in a coastal basin.

Consequently, the use of mathematical simulation is of great significance for the diagnosis as well as the prognosis and prevention of hazardous situations that may threaten the marine environment. From this point of view, the description and application of three different types of models, presented in the following sections, constitutes the objective of the present chapter.

2. Methods and case studies

2.1. Trix — Estimating the trophic status of a coastal ecosystem

2.1.1. Description of the method

The trophic index (TRIX) proposed by Volleinweider et al. [7] is estimated as a linear combination of the logarithms of four variables, referring to primary production [chlorophyll-a (Chl-a), dissolved inorganic nitrogen (DIN), total phosphorus (TP) and the absolute percentage deviation from oxygen saturation (aD%O)], in the form:

$$TRIX = \frac{\log\left(Chl - a * aD\%O * DIN * TP\right) + k}{m} \tag{1}$$

Using data from the Northern Adriatic Sea, Volleinweider et al. [7] suggested the following values for the parameters k and m: k=1.5 (to fix the differences between the upper and lower range of the variables to 3 log units) and m=1.2 (to fix the scale range from 0 to 10). Low TRIX values (2-4) indicate poor productivity in waters, and thus imply "high quality" waters, while TRIX values from 4 to 5 indicate "good quality", 5 to 6, "mediocre quality" and 6 to 8, "bad quality" waters [8].

In this form, TRIX has been applied for the classification of coastal waters in the Mediterranean Sea, the Marmara Sea, the Northern European seas, the Black Sea, the Caspian Sea, the Baltic coastal waters, the seas of Southeast Mexico, and the Persian Gulf [9-17]. However, when the concentrations of the variables appear with different upper and lower ranges in respect to those of the Northern Adriatic Sea, the TRIX index needs modification by re-estimating the k and m parameters according to the new upper and lower limits [7, 18].

An unscaled trophic index, UNTRIX, was introduced by Pettine et al. [10] following the European Water Framework Directive (WFD, 2000/60/EC):

$$UNTRIX = \log\left(Chl - a * aD\%O * DIN * TP\right) \tag{2}$$

http://dx.doi.org/10.5772/59498

Two classification procedures for water were proposed. The first is based on a simple comparison between box and whisker plots for UNTRIX data from: 1) the investigation site and 2) the respective data from the reference area. The second procedure is based on the TQR_{TRIX} trophic index, defined according to equation (3):

$$TQR_{TRIX} = \frac{50^{th}\,UNTRIX_{reference}}{75^{th}\,UNTRIX_{site}} \tag{3}$$

where 50th is the median and 75th *percentile* is a value indicating that 75 % of the values of the variable fall below that value.

2.1.2. Case study

The three abovementioned trophic indices (TRIX, UNTRIX and modified TRIX) were applied to quantify the water quality of the coastal waters in the Kalamitsi area, Ionian Sea, Greece (Fig. 1), before and after the operation of the Waste Water Treatment Plant (WWTP) [11]. As mentioned, this quantification allows a classification of the water quality.

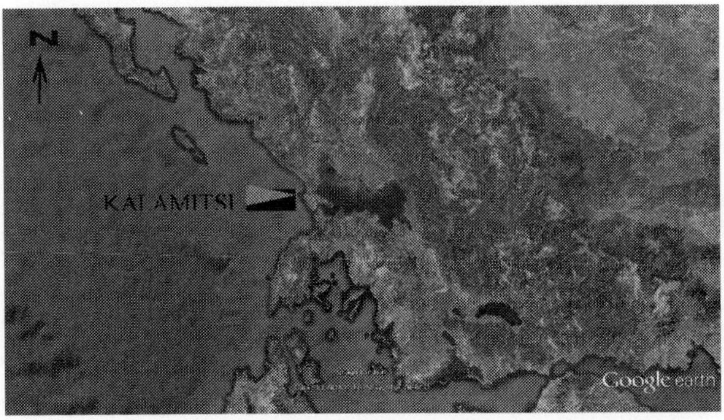

Figure 1. The study area of Kalamitsi, Ionian Sea, Greece (Google Earth)

Data were collected seasonally, from three fixed stations, from March 2001 until January 2003 (before WWTP operation) and from August 2004 until July 2006 (after WWTP operation).

Before the WWPT operation, the TRIX values showed a temporal variation in the range of 1.9-4.7 TRIX units, while no spatial variation (between stations) was observed. The mean TRIX value was 3.5. After the WWPT operation, the TRIX values showed temporal and spatial variations (2.2-4.8 TRIX units). The mean TRIX value was 3.4. As can be seen, no significant differences in TRIX values were detected before and after the WWPT operation (T-test, p=0.608). These values indicate "high" water quality, in contrast with the TRIX values obtained for the gulfs of the Aegean Sea, like the Thermaikos Gulf (5.0-6.0) and the Saronikos Gulf (3.7-6.2), with a mean of 5.3, indicating "mediocre" water quality [14].

As the concentrations of Chl-a and DIN from Kalamitsi displayed a profile much lower than those of the Northern Adriatic Sea [11, 7], the TRIX was re-estimated (KALTRIX) according to the Kalamitsi data set, following this procedure:

- zero values of the variables were substituted by the detection limit value of the applied analytical method
- extreme values (m ± 2.5 SD) were excluded (where m is the mean value and SD is the standard deviation of the sample)
- the range between upper and lower values of each variable was set to 1 log unit
- test of normality was performed for log transformations.

This procedure furnished equation (4):

$$KALTRIX = 2.5 * \log\left(Chl - a * aD\%O * DIN * TP\right) - \left(0.5\right) \tag{4}$$

KALTRIX values displayed high temporal variability (0.8-9 units) before the WWPT operation, whereas both temporal and spatial variability occurred after the WWPT operation. In both cases, KALTRIX values were higher than those of TRIX, although the same data set was applied. The mean KALTRIX value was 5.3 before WWPT operation and 5.0 after it, with no statistically significant differences (T-test, p=0.735), indicating minor influence on water quality of WWTP operation. However, greater divergences of KALTRIX values were observed after WWTP operation. So, it could be assumed that KALTRIX is more sensitive, and is able to discriminate minor trends not easily detected by TRIX.

The same data set from Kalamitsi and data from South Evoikos used as a reference site (which has been reported as a having high-good ecological quality status, directive 2000/60/EC) were applied to the UNTRIX index (equation 2). Using the box and whisker plots (Table 1), we defined the 100th, 75th and 50th plots of the Southern Evoikos and Kalamitsi areas before and after the WWPT operation.

	S. Evoikos (reference site)	**Kalamitsi** (before WWTP operation)	**Kalamitsi** (after WWTP operation)
100th	3.097	4.091	4.063
75th	2.118	3.142	2.994
50th	1.768	2.673	2.452

Table 1. The 100th, 75th and 50th box and whisker plots of UNTRIX values from S. Evoikos (reference site) and Kalamitsi before and after the WWTP operation

According to the TQR_{TRIX} index (equation 3), water quality is classified as "moderate" both before (0.56) and after (0.59) the WWTP operation. On the other hand, according to the box-plot procedure [10], water quality is classified as "moderate" before the WWTP operation (75th of the site = 3.142>3.097=100th of the reference), while after the WWTP operation it is classified as "good" [75th of the Kalamitsi =2.994>2.118=75th of the reference (S. Evoikos) and simultaneously 100th of the Kalamitsi = 4.063>3.097=100th of the reference (S. Evoikos)].

2.2. Piecewise linear regression models with breakpoints

2.2.1. Description of the method

If an investigation is focused not only on the classification of the coastal waters' quality but also on the assessment of critical thresholds, then piecewise linear regression models with breakpoints may constitute useful tools [19]. Furthermore, piecewise linear regression models with breakpoints are able to express discontinuity appearing in ecosystems, associated either with habitat fragmentations (usually independent variables) or with "decisions" taken by individuals according to their situation [20, 21]. The breakpoint (bt) is a value of either an independent variable (x) or a dependent variable (y) according to the specific characteristics of the ecosystems and the interests of the investigation, where two separate linear regression equations are estimated: one with y values less or equal to bt and the other one with y values greater than bt. The general form is given by equation 5:

$$y = \begin{cases} b_{01} + b_{11}x_1 + \ldots + b_{n1}x_n & \text{for x or y} \le \text{bt} \\ b_{02} + b_{12}x_1 + \ldots + b_{n2}x_n & \text{for x or y} > \text{bt} \end{cases} \qquad (5)$$

Such models have been used, for example, to estimate the production of crops, to search for thresholds in coastal mangrove forests and to examine a limpet feeding rate and its response to temperature [22-24].

2.2.2. Case study

As chlorophyll-a and nutrients are two of the most common indicators for eutrophication assessment [25], a piecewise linear regression model was applied with data (275 observations) for chlorophyll-a and nitrates (NO_3-N) from the Thermaikos Gulf, Greece, which were collected over a period of two years [20].

Nitrogen was chosen because it is used by cells for the biosynthesis of molecules like chlorophyll-a, DNA, RNA and proteins. Thus, it could be assumed that discontinuity might have appeared because each phytoplankton individual "decides" the way that a given amount of nitrogen will be used, depending on the stage of its cell cycle and cell demands. In the fitted model, the concentration of dissolved nitrogen (NO_3-N, μg/l) is the independent variable, while the concentration of chlorophyll-a (mg/m^3) is the dependent one.

$$Chl - a = \begin{cases} a_1 + b_1 * NO_3 & \text{if Chl-a} \le \text{bt} \\ a_2 + b_2 * NO_3 & \text{if Chl-a} > \text{bt} \end{cases} \qquad (6)$$

Because the above model is piecewise linear, the parameters were estimated using the quasi-Newton method, an interactive method that minimizes the least squares loss function [Σ (observed-predicted)^2] through iterative convergence of the predefined empirical equation.

According to the present data, the breakpoint that minimizes the loss function and simultaneously gives the highest correlation coefficient (r=0.84) is 2.5 mg/m^3 of chlorophyll-a. So the model of equation (6) assumes the form:

$$Chl-a=\begin{cases}1.3787-0.08441*NO_3 & \text{if Chl-a} \le 2.5\\ 4.39438-0.16438*NO_3 & \text{if Chl-a} > 2.5\end{cases}$$

For all the estimated parameters, the t-test was implemented with:

$$H_0\text{: parameter} = 0$$
$$H_1\text{: parameter} \neq 0$$

In all cases, H_0 is rejected as $p<0.05$. The fitted model explains about 71 % of the observed variability ($r^2=0.706$).

Plotting the predicted and observed values [20], it was found that the predicted values of Chl-a were overestimated when observed values were lower than 0.9 (mg/m^3), and underestimated when observed values were higher than 4.4 (mg/m^3). Thus, we simulated the model starting with a lower breakpoint value (0.5 mg/m^3 Chl-a) and terminated it at breakpoint = 4.8 mg/m^3. The results show that when breakpoint values are lower than 0.9 or higher than 4.4 mg/m^3 Chl-a, no relationship could be estimated between Chl-a and NO_3. However, when 0.9≤breakpoint≤4.4 mg/m^3 Chl-a, a relationship between both parameters appeared. The negative slopes b_i (-0.08411 and -0.16438 respectively) indicate a "tendency of the ecosystem to be stabilized" at the values of 2.5 or 0.9 mg/m^3 of Chl-a, when its concentration in the Thermaikos Gulf varies between 2.5 and 4.4, and 0.9 and 2.5, respectively. It can be assumed that, when Chl-a values are higher than 2.5, then competition and predation may tend to stabilize the phytoplankton community, while processes like decomposition and inflows may increase NO_3 concentration at the same time.

The estimated critical values of breakpoints (0.9, 2.5 and 4.4) are very close to the criteria for trophic assessment in marine and coastal ecosystems given by the Swedish Environmental Protection Agency (**<1.5**, 1.5-2.2, **2.2-3.2**, 3.2-**5** and >5 μg/l Chl-a) and the low risk "trigger values" (2-4 μg/l Chl-a) which were proposed by the Australian and New Zealand Environment and Conservation Council [26]. So, it could be assumed that, in the case of the Thermaikos Gulf, values of Chl-a lower than 0.9 may indicate oligotrophic waters, from 0.9 to 2.5, mesotrophic, from 2.5 to 4.4, eutrophic, and above 4.4, hypertrophic waters. Furthermore, it has been reported that Chl-a concentrations between 2 and 4.44 appeared during algal blooms [27,28], so it could be suggested that values of Chl-a close to the breakpoint may indicate the start of an algal bloom, although more field and laboratory investigations are needed to confirm this hypothesis.

2.3. A numerical dynamic hydrobiological model

2.3.1. Description of the method

In this part of the chapter, a hydrobiological/bio-hydrodynamic model is described. This model is constituted by two parts: a first hydromechanical part, which computes the hydrodynamics and the matter transfer in a coastal geophysical basin; and a second hydrobiological part, which computes the cells' growth rate (generation of new mass) and the cells' decay (loss of mass). When shallow waters characterize a coastal basin, a two-dimensional, depth-averaged, hydrodynamic model is considered to be quite sufficient for the simulation of the seawater circulation. In this case, the hydrobiological model is also a depth-averaged model, like the one presented below.

As mentioned above, **the first part of the model refers to the hydromechanical processes.** These hydromechanical processes concern (a) the hydrodynamics, e.g., the computation of the velocity field and the sea surface elevation field; and (b) the matter transport, described by the advection and dispersion processes.

Hydrodynamics. The hydrodynamic-hydromechanical model is based on the usual equations for the conservation of mass and momentum, as described in equations (7) [29, 30, 6]:

$$\begin{aligned}
&\frac{\partial U}{\partial t}+U\frac{\partial U}{\partial x}+V\frac{\partial U}{\partial y}=-g\frac{\partial \zeta}{\partial x}+fV+\frac{\tau_{sx}}{\rho\, h}-\frac{\tau_{bx}}{\rho\, h}+v_h\frac{\partial^2 U}{\partial x^2}+v_h\frac{\partial^2 U}{\partial y^2} && a\\
&\frac{\partial V}{\partial t}+U\frac{\partial V}{\partial x}+V\frac{\partial V}{\partial y}=-g\frac{\partial \zeta}{\partial y}-fU+\frac{\tau_{sy}}{\rho\, h}-\frac{\tau_{by}}{\rho\, h}+v_h\frac{\partial^2 V}{\partial x^2}+v_h\frac{\partial^2 V}{\partial y^2} && b\\
&\frac{\partial \zeta}{\partial t}+\frac{\partial (Uh)}{\partial x}+\frac{\partial (Vh)}{\partial y}=0 && c
\end{aligned} \tag{7}$$

where h is the seawater depth, U and V are the vertically averaged horizontal velocities, ζ is the sea surface elevation, f is the Coriolis parameter, τ_{sx} and τ_{sy} are the wind surface shear stresses, τ_{bx} and τ_{by} are the bottom shear stresses, ν_h is the dispersion coefficient related to the local vorticity, ϱ is the density of the water and g is the gravity acceleration [29, 31].

The output from the hydrodynamic simulation, e.g., the velocity field, is then used as an input in the transport simulation. The specific transport model applied here is based on the tracer method which can be described as a Lagrange-Monte Carlo simulation (random walk method) [29, 3]. The simulation of the transport processes based on this method has the important advantage of avoiding numerical problems. Advection and dispersion processes are simulated by means of this method [32-36]. In this specific work, the two parts of the hydromechanical model, i.e., the hydrodynamic and transport processes, are fully coupled, including the biological processes at the same time. In more detail, the transport part of the model uses, as already mentioned, the velocity field, produced by the hydrodynamic model at each time step. The transport model is described analytically in the following paragraphs.

Matter transport. According to the tracer method, a large number of particles that represents a particular amount of mass are released in the marine environment through a source. The location where an algal bloom episode occurs determines the position of that source. The transport and fate of each particle is traced with time. Advection of the particulate matter is computed by the local seawater velocity while turbulent diffusion is simulated by the random Brownian motion of the particles. More specifically, the tracer method is described by the following steps [29, 6]:

a. the velocity field is determined by a set of values of the velocity components at specific grid points computed by the hydrodynamic part of the model

b. a suitable time step is selected

c. the range of the random velocity ± Ur is computed from the following equation (8):

$$U_r = \sqrt{6D/dt} \tag{8}$$

where D is the local diffusion coefficient and dt is the time step

d. in the case of instantaneous discharge, a specific amount of particles is released at once from the initial source; while in the case of continuous discharge, for each time step a specific number of particles is introduced in the study domain

e. integration in time is executed and the new coordinates of each particle are computed; the motion of each particle is analysed considering: i) a deterministic part which concerns the advective transport, and ii) a stochastic part which concerns the transport-spreading due to diffusion processes.

The horizontal positions of the particles are computed from the superposition of the deterministic and stochastic displacements:

$$x_i^{n+1} = x_i^n + \Delta x_i^n + \Delta x_i^{n\,\prime} \text{ with } \Delta x_i^n = u_i^n\left(x_i^n, t^n\right) dt \text{ and } \Delta x_i^{n\,\prime} = u_i^{n'}\, dt\; rnd[-1,1] \tag{9}$$

$$y_i^{n+1} = y_i^n + \Delta y_i^n + \Delta y_i^{n\,\prime} \text{ with } \Delta y_i^n = v_i^n\left(y_i^n, t^n\right) dt \text{ and } \Delta y_i^{n\,\prime} = v_i^{n'}\, dt\; rnd[-1,1] \tag{10}$$

where Δx_i^n and Δy_i^n are the deterministic displacements and $\Delta x_i^{n'}$ and $\Delta y_i^{n'}$ are the stochastic displacements, with u_i^n (x_i^n, t^n) and v_i^n (y_i^n, t^n) being the deterministic velocities at time t^n at the location x_i^n and y_i^n of the i particle, $u_i^{n'}$ & $v_i^{n'}$ are the random (stochastic) horizontal velocities at time t^n at the locations x_i and y_i, respectively, $u_i^{n'} = v_i^{n'} = \sqrt{(6D_h/dt)}$, with D_h being the horizontal particle diffusion coefficient, and rnd is a random variable distributed uniformly between -1 and +1. Eventually, the spatial particle distribution allows for the computation of the particle concentrations in each grid box.

The second part of the model refers to the hydrobiological processes

If we define

- N_0 as the starting population of a phytoplanktonic species
- μ as the growth rate (divisions/day) and
- TL as the total losses

then these three factors regulate the density of population N, which, in a continuous model, follows an exponential growth, according to equation (11):

$$N = N_0 \cdot e^{(\mu - TL)t} \tag{11}$$

where t is the time (in days), TL is influenced by various processes such as predation, sinking, water movement and stratification, and μ is dependent on abiotic factors such as temperature, light and nutrients, since they act directly on the individuals' biochemistry [37, 38].

In the biological part of the model, the maximum growth rate of phytoplankton $\mu_{(T)}$ is considered to follow Eppley's equation [39, 40]. Then light intensity and sun shine duration are considered to limit $\mu_{(T)}$ according to Steele's equation [41]. The sunshine duration and the photosynthetic active radiation (par) were estimated using field data [42, 43]. Nitrogen and phosphorus limitations (N_{lim} and P_{lim} respectively) are computed according to Michaelis-Menten enzyme Kinetics. Temperature and nutrients concentrations are considered to take random values, following normal distribution which was estimated from field data [27].

The abovementioned processes, with the relevant factors as functions of different elements and the relevant constants, are given in Tables 2a and 2b.

Function	Definition	Formula	References
μ	Phytoplankton growth rate (day^{-1})	$\mu_{(T)} . L_{lim} . N_{lim} . P_{lim}$	[44, 37]
$\mu_{(T)}$	Temperature-dependent growth rate (°C)	$0.81\ e^{0.0631\ T}$	[39, 40]
L_{lim}	Light limitation	$\frac{I}{I_{opt}} \cdot e^{1-\frac{I}{I_{opt}}}$	[41]
N_{lim}	Nitrogen limitation	$\frac{N_t}{K_N + N_t}$ (Nt: μmole N)	Michaelis–Menten enzyme kinetics
P_{lim}	Phosphorus limitation	$\frac{P_t}{K_P + P_t}$ (Pt: μmole P)	Michaelis–Menten enzyme kinetics

(a)

Constants	Definition	In model	In literature
TL	Total losses (grazing + sinking + natural death)	~ 0.33 (day-1) estimated from numerical tests and bibliographical references	0.07-0.67 (season dependent) [45-48]
I_{opt}	Optimum light irradiance	110 W. m^{-2}	100-170 W. m^{-2} [37, 49]
K_N	Half-saturation for nitrogen uptake	3 μM	0.4-10 μM [44, 49-51]
K_P	Half-saturation for phosphorus uptake	0.2 μM	0.09-3.4 μM [52-54]

(b)

Table 2. (a) processes used in the biological compartment of the model (from: [6]) b) constants used in the relations of the previous table (from: [6])

In more detail, the information for the growth rate coefficients was introduced into the model as a time series of mean daily growth rates. That is, it was computed in a discrete way, instead of using the continuous equation (11). The modelling approach, concerning the increase of the particles due to division and the decrease due to particle losses, is described by the following steps [6]:

a. After a time period of one day, a new number of particles equal to ($\mu \bullet N_0$) are generated. In that product (i.e. $\mu \bullet N_0$) μ is the growth rate coefficient and N_0 is the number of particles before the division, denoting the total number of particles of the previous day. The "new" total number of particles N, after the growth process, is afterward computed from the equation $N=N_0 + \mu \bullet N_0$.

b. The new particles' position is determined from the positions of other particles randomly selected.

c. Additionally, a decay coefficient (TL) is adopted for the population decrease and particle disappearance of the water column. So, at the end of a period of a day, a number of particles equal to $TL \bullet N$ is drawn out, where TL is the aforementioned decay coefficient and N is the total number of particles of the previous day, plus the particles added due to the growth process of the present day, e.g., $TL \bullet N = TL \bullet (N_0+\mu \bullet N_0)$. These particles are no longer taken into account in the computations.

d. Finally, the total number of particles Nt, after growth and decay process, is given by eq. 13:

$$N_t = N_0 + \mu \cdot N_0 - \left[TL \cdot \left(N_0 + \mu \cdot N_0 \right) \right] \Rightarrow N_t = N_0 (1 + \mu - TL - \mu \cdot TL) \tag{12}$$

Concerning the boundary conditions for the transport model, it should be pointed out that if the particles reach the coast, they return to their previous position, while the particles reaching the open sea boundary are trapped there and no further computation is made for them; these particles are then excluded from the computational loops. The position of the source is determined from the place where an episode of an algal bloom is observed. Particles' concentrations are then computed from the number of particles, counted at each grid box. As far as the initial conditions are concerned, the numerical simulations always start from scratch, i.e., zero current velocities at the starting point of the simulation time.

2.3.2. *Case studies*

The hydrobiological model

The application of the mathematical hydrobiological model described in the previous section is presented here.

a. **Application to a real episode in the Thermaikos Gulf**. The model runs were based on meteorological data for a particular period in which a real episode of algal bloom took place in the Thermaikos Gulf, starting from January 10th, 2000. The Thermaikos Gulf is a semi-enclosed coastal area, located in the north-western corner of the Aegean Sea, in the east of the Mediterranean Sea, as depicted in Figs. 2 and 3. The Axios, Aliakmon and Loudias are the three main rivers that discharge into the western and north-western coasts of the gulf, while the River Pinios outflows much further south, in the south-western coast of the outer gulf. During recent decades, there has been a major reduction in the total freshwater input to the gulf, mainly as the result of the extraction of river water for irrigation, and the construction of a series of hydroelectric power dams on the Aliakmon [55, 56].

The hydrodynamic model was fully coupled with the transport and biological models described in the previous sections. The equations of the hydrodynamic model were numerically solved by the finite difference method on an orthogonal, staggered grid (Arakawa C grid), where U and V components refer to the nodes' sides and ζ refers to the interior of each mesh. The field of the gulf was discretized using 32×36 grid cells. The spatial-discretization step was dx = 1000 m.

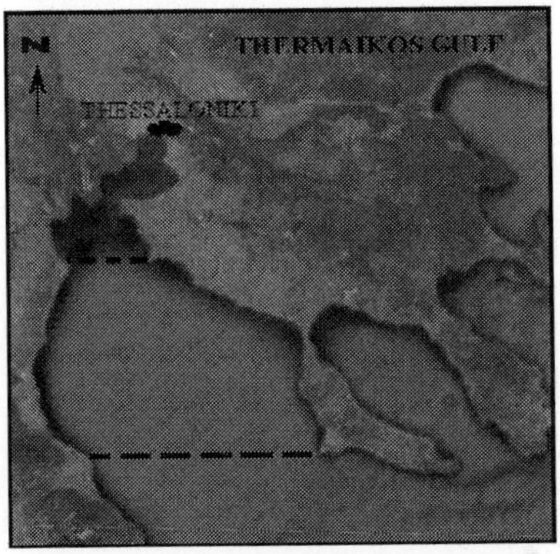

Figure 2. The Thermaikos Gulf (extending from Thessaloniki's coastline to the upper dash line) and the extended basin of the gulf (extending from Thessaloniki's coastline to the lower dash line) (Google Earth)

Fig. 3a depicts the study area of the Thermaikos basin, with A being the initial location of the appearance of an HAB episode, B the control position-location with reported measured values of HAB concentrations, T the city of Thessaloniki, P the port of Thessaloniki, E the area called Megalo Emvolo, and M1, M2 and M3 the mussel culture areas lying at the west coasts of the gulf. Fig. 3b depicts the grid described above, as the platform for the numerical solution of the equations of the model.

The present application to the real coastal basin of Thermaikos Gulf tried to simulate the fate of harmful algal cells after a sudden appearance of HAB, using *Dinophysis* abundances 10^4-10^5 cells/lit recorded in north-eastern Thermaikos on January 10th, 2000 (position A in Fig. 3a). For the simulation, a number of 1000 particles was initially used at the location (A), and 619,000 particles were finally counted 21 days later (this latter number included particles that may have dispersed in the coastal basin of the gulf or escaped out of the gulf). The model runs led ultimately to the distribution of *Dinophysis* populations in the coastal basin of the gulf. Two

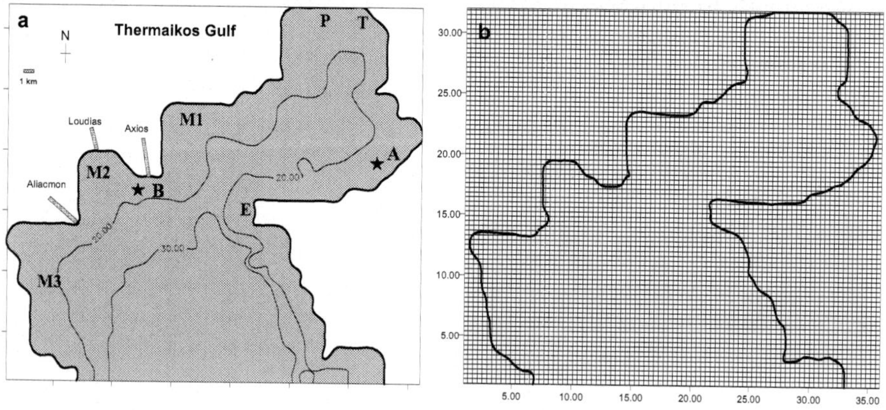

Figure 3. (a) Thermaikos Gulf, where A and B are control positions for the model, T the city of Thessaloniki, P the port of Thessaloniki, E the area called Megalo Emvolo, M1, M2 and M3 mussel culture areas and (b) the grid of the basin's discretization. The axes' coordinates are in number of spatial steps (from: [6]).

control points were used corresponding to the locations A and B, as depicted in Fig. 2. The hydrodynamic circulation of the gulf was activated by variable winds prevailing over this period (between January 10th and 31st, 2000, with a predominance of northern winds that correspond to the most frequent conditions in the Thermaikos Gulf). Concerning the effect of tide on the general pattern of dispersion, it is considered negligible, since the mean tidal signal of 25 cm is quite small and the residual currents due to the tide are negligible. Moreover, riverine water inputs and seawater density differences in general were not taken into account, because the total freshwater input of the rivers has been dramatically reduced in the last decades. Finally, the present numerical simulation was performed for wind-generated circulation, under the winds blowing over the area during the period between January 10th and 31st, 2000. The time series of wind data was recorded from a meteorological station of the Institute of Forestry Research at the location of Sani, Chalkidiki. As mentioned, the transport model was applied considering the position A as the local source of particulate matter and instantaneous appearance of algal bloom, where large concentrations of *Dinophysis* (HAB) were recorded in the field.

The daily variable estimated growth rate used in the biological model has a range between 0.176 and 0.489 divisions per day [6]. These values are comparable with those estimated using the long-term (one year) deterministic model for dinoflagellates in the same area [37, 38]. In the literature, the values of dinoflagellates' growth rates vary within a range of 0.04-0.7 divisions per day, depending on time and local conditions [1, 48, 57].

The model results concerning the distribution of the harmful phytoplankton cells for time periods of five days, 10 days and 21 days after the most intense bloom of January 10th are given in Figs. 4a, 4b and 4c [6]. According to the abovementioned patterns of the *Dinophysis* dispersion, it may be observed that five days after the bloom at location A (Fig. 4a), *Dinophy-*

http://dx.doi.org/10.5772/59498

sis masses were transported and concentrated in the area of Megalo Emvolo (position E in Fig. 2) west of position A; 10 days after the initial bloom (Fig. 4b), higher concentrations of *Dinophysis* reached some coastal areas of the west and north-west Thermaikos Gulf, where the largest mussel farms of Greece lie; finally, 21 days after the bloom (Fig. 4c), the present results show that concentrations of *Dinophysis* have been dispersed to a relatively large area of the outer Thermaikos Gulf, and mainly along the west coast of the gulf [6]. Of course, because a large part of the western as well as the eastern coastline consists of bathing coasts, e.g., tourist sea beaches, situations of large algal concentrations are not only aesthetically undesirable but also hazardous on some occasions.

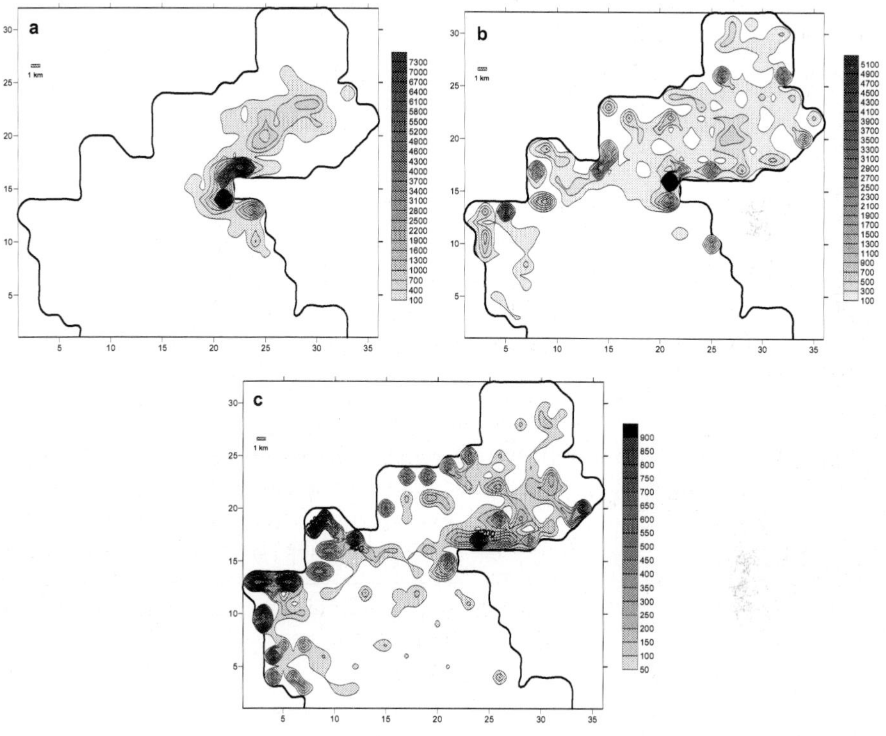

Figure 4. The *Dinophysis* concentrations (a) five days, (b) 10 days and (c) 21 days after the bloom in position A (cells/lt). The axes' coordinates are in a number of spatial steps (from: [6]).

In more detail, the simulation shows that 21 days after the bloom at the north-eastern side of the inner Thermaikos Gulf, with an initial concentration of 57,000 cells/lit at position A, and under the influence of prevailing northern and north-eastern winds, the main mass was spread to the central and eastern coasts of the inner Thermaikos, as well as to the western coasts of the gulf. Concentrations higher than 200 cells/lt were computed for the region north-east of the Aliacmon river-mouth and south of the Axios river-mouth (position B in Fig. 2). These concentrations are very close to those reported by [27]. Concerning the north-western coasts of the gulf, e.g., the region of mussel cultures north-east of the Axios delta (north of position B, area M1), the model resulted to values of up to 400 cells/lt. Furthermore, concerning the west and south-western coasts of the gulf, hosting the mussel cultures south of the Loudias river-mouth (mussel culture area M2, Fig. 2) and west and south of the Aliakmon river-mouth (mussel culture area M3, fig. 2), the of the model resulted to values of up to 900 cells/lt. The *Dinophysis* concentrations reached similar values at the northern area of Megalo Emvolo (position E, fig. 2). Along the coast of Thessaloniki City (position T, Fig. 2), significant *Dinophysis* masses were not found [6].

b. Application of different scenarios to the extended area of the Thermaikos basin

The application of the aforementioned hydrobiological model, based on different wind conditions prevailing over the extended area of the Thermaikos Gulf, is presented in this section. To this end, the outer part of the external Thermaikos Gulf was also included in the present simulations (Fig. 6a). Furthermore, this application was focused on the investigation of the critical conditions for the appearance of an algal bloom episode and the dispersion of its population in space and time. Additionally, different positions in the Thermaikos Gulf, as sources of the initial population, and different starting population densities were examined.

The equations of the hydrodynamic model, in both aforementioned cases, were numerically solved by the finite difference method, on an orthogonal staggered grid (Arakawa C grid) similar to the previous bathymetry (without the greater area of the outer Thermaikos Gulf). The study area was now discretized with a grid of 41×42 cells, with spatial step dx = 1852 m (Fig. 5b).

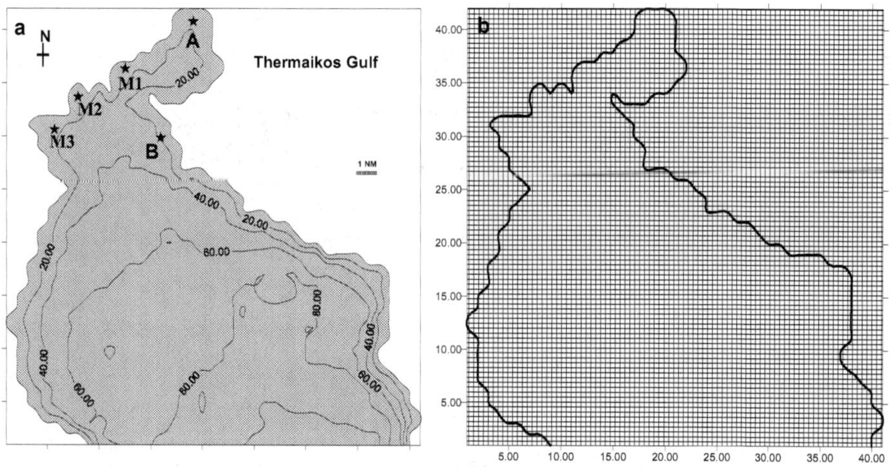

Figure 5. (a) Thermaikos Gulf, Greece. A and B are the starting points of the simulation, A is the area close to the harbour of Thessaloniki, M1, M2 and M3 are mussel cultures areas [5], and (b) is the grid of the basin's discretization. The axes' coordinates are in number of spatial steps.

http://dx.doi.org/10.5772/59498

In this simulation, the following two assumptions drawn from the literature were taken into account:

a. Although the cell concentration (or abundance) in the peak of the bloom exceeds the number of 10^4 cells/lt, a minimum of 500 to 1200 cells/lt is a threshold for restrictions in fisheries [44]. So, in this study, 2000/lt cells of *Dinophysis spp.* were considered to be the minimum population for starting a bloom episode.

b. In almost all the studies, before a bloom episode, *Dinophysis* cell abundance is usually less than 200 cells/lt. For this reason, we consider here $N_o = 150$ cells/lt.

The initial model runs' simulation started with a low phytoplankton net growth rate of $\mu = 0.3$ divisions per day. Finally, the simulation tests showed that HAB episodes appeared when the net growth rate takes values close to 1.0 divisions per day, while the losses were embodied in the net growth rate.

A number of 1000 particles was initially used at the location A, and 128,000 particles were finally counted at the end of the simulation. The total simulation time was taken as equal to seven days (time period of one week) which was sufficient for the flow to reach steady state conditions (either for N or for S winds, with speeds of 2 and 10 m/s). The patterns of seawater circulation, corresponding to steady state flow, under the influence of north and south winds of 10 m/s, are given indicatively in Fig. 6.

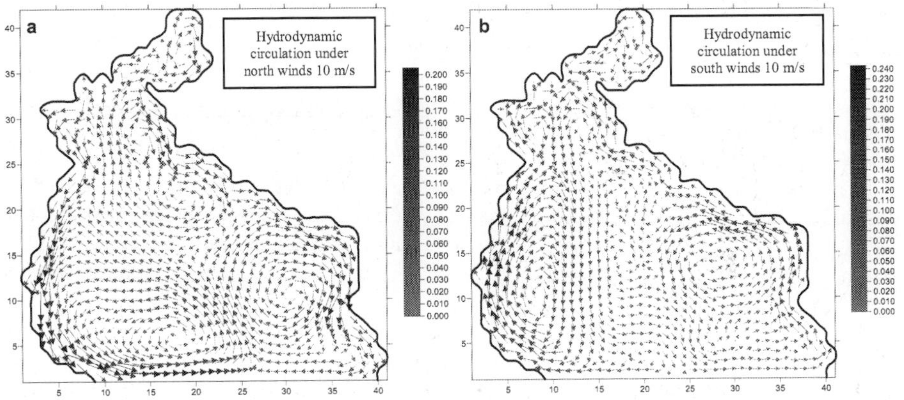

Figure 6. Current velocities (m/s) under the influence of (a) N winds (left) and (b) S winds, 10 m/s (right). The axes' coordinates are in number of spatial steps.

Finally, the combination of the following cases was considered:

a. two different positions for the starting point, the first one, position A, in the inner part of the gulf, and the second one, position B, in the outer gulf, as depicted in Fig. 5a,

b. two different wind directions, the north (N) and the south (S) winds, which are generally the most frequent in the study area of the gulf,

c. two different wind speeds, the first one corresponding to low winds of 2 m/s and the second one corresponding to relatively strong winds of 10 m/s.

These scenarios led to eight different patterns of algal dispersion, as presented in the following sequence: four patterns corresponding to the case of initial algal bloom at location A, and four patterns corresponding to the case of initial algal bloom at location B (with A and B depicted in Fig. 5a).

Algal bloom at location A

More specifically, in the case of starting an algal bloom at location A in the inner part of the gulf, the patterns of the dispersion of the cells under the influence of north winds with speeds of 2 and 10 m/s, respectively, are given in Fig. 7; the patterns of the dispersion of the cells under the influence of south winds with speeds of 2 and 10 m/s, respectively, are given in Fig. 8.

The following characteristics [5] can be described:

- If the source is located in the Thessaloniki Gulf (inner part of the Thermaikos Gulf), the dispersion of the bloom in the outer part of the Thermaikos Gulf is rather towards the rear
- A bloom can be started with a low population (150 cells/lt) and a net growth rate close to 1.0 divisions per day, in a period of one week
- South winds with a low speed of 2 m/s can cause higher cell concentrations than north winds with the same speed.

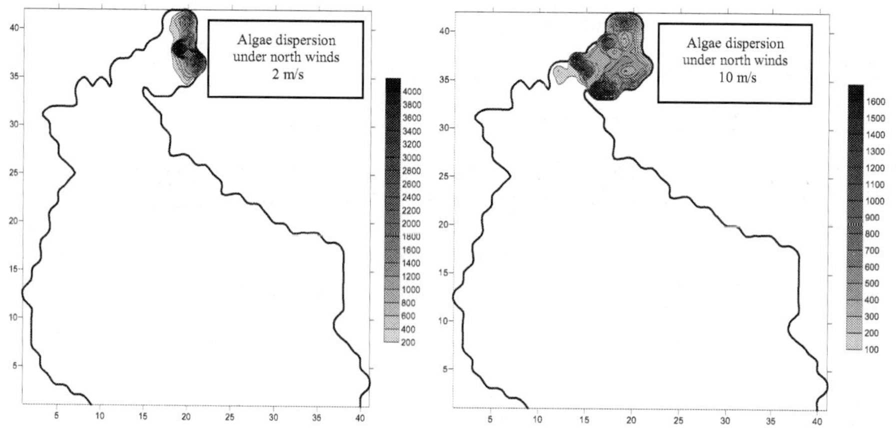

Figure 7. The concentrations of algae (cells/lt) one week after the bloom in position A under the influence of north winds with speeds of 2 m/s (left) and 10 m/s (right) (from: [5]). The axes' coordinates are in number of spatial steps.

Algal bloom at location B

In the case of starting an algal bloom at location B in the outer part of the gulf, the patterns of the dispersion of the cells under the influence of north winds with wind speeds of 2 and 10 m/

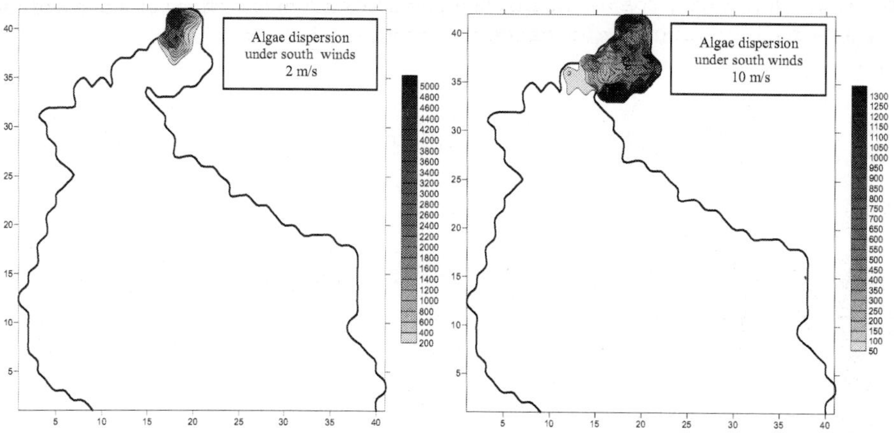

Figure 8. The concentrations of algae (cells/lt) one week after the bloom in position A under the influence of south winds with speeds of 2 m/s (left) and 10 m/s (right) (from: [5]). The axes' coordinates are in number of spatial steps.

s, respectively, are given in Fig. 9; the patterns of the dispersion of the cells under the influence of south winds with wind speeds of 2 and 10 m/s, respectively, are given in Fig. 10.

The following characteristics [5] can be described:

- An algal bloom may appear only when wind speed is low (~2 m/s)
- South winds may cause dispersion of phytoplankton cells in the inner gulf.

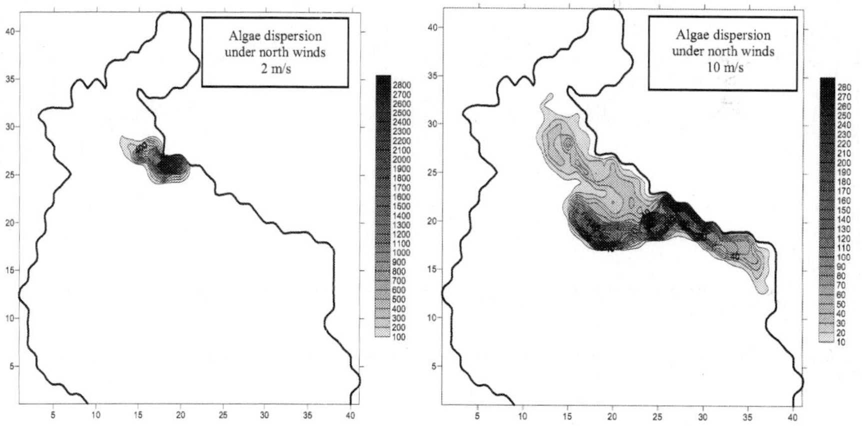

Figure 9. The concentrations of algae (cells/lt) one week after the bloom in position B under the influence of north winds with speeds of 2 m/s (left) and 10 m/s (right) (from: [5]). The axes' coordinates are in number of spatial steps.

More analytically, when the source is located at position B in the outer gulf, north winds do not seem to cause transport and dispersion of phytoplankton cells at the inner gulf during the period of one week. However, cells can be found along the eastern coasts of the outer gulf. In this case (source at position B), HABs were not observed in the present calculations, except under the influence of low winds (~2 m/s). South winds may transfer and disperse phytoplankton cells in the inner gulf and close to the areas of mussel culture after a period of one week, but phytoplankton concentrations in all cases are shown to be lower than 1000 cells/lt.

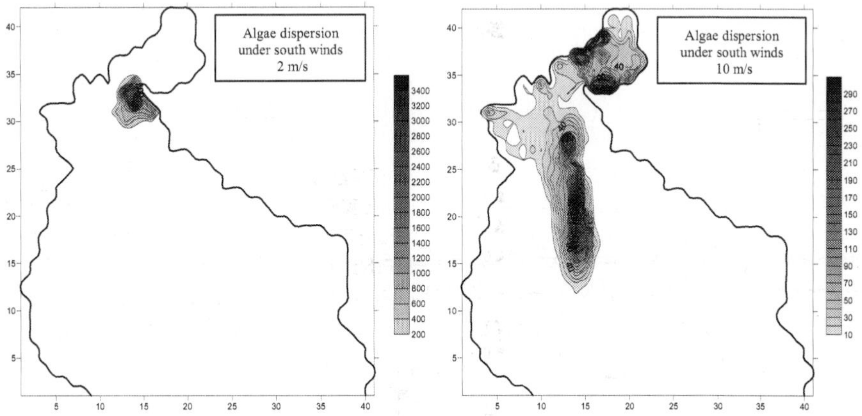

Figure 10. The concentrations of algae (cells/lt) one week after the bloom in position B under the influence of south winds with speeds of 2 m/s (left) and 10 m/s (right) (from: [5]). The axes' coordinates are in number of spatial steps.

2.4. A stochastic markovian model

2.4.1. Model description

As randomness and uncertainty appear in natural ecosystems, Markovian models, as a part of stochastic processes, have been applied in many ecological processes, including succession, population dynamics, energy flow and diversity [37, 58-61]. In Markovian models, the definition of "states" is a critical point and depends on the specific characteristics of the system to be modelled. A state consists of individuals, variables or number of species, while all the possible states of the system correspond to the "state space" C (C={1,2,3...,k}). The vector $n(t)=[n_1(t), n_2(t),..., n_k(t)]$ gives the expected number of the members on each one of the k "states", for each t (Table 1). The so-called transition probability ($p_{ij}(t)$) is the probability of a member moving from state *i*, at time *t*, to state *j* at time *t+1*. Transition probabilities can be estimated as:

$$p_{ij} = \frac{n_{ij}(t+1)}{n_i(t)} \text{ where, } i,j = 1,2,3,...k$$

To states		To states at time (t+1)			
From states	time (t)	1	2	..	k
1	$n_1(t)$	$n_{11}(t+1)$ $p_{11}=\frac{n_{11}(t+1)}{n_1(t)}$	$n_{12}(t+1)$ $p_{12}=\frac{n_{12}(t+1)}{n_1(t)}$	..	$n_{1k}(t+1)$ $p_{1k}=\frac{n_{1k}(t+1)}{n_1(t)}$
2	$n_2(t)$	$n_{21}(t+1)$ $p_{21}=\frac{n_{21}(t+1)}{n_2(t)}$	$n_{22}(t+1)$ $p_{22}=\frac{n_{22}(t+1)}{n_2(t)}$	..	$n_{2k}(t+1)$ $p_{2k}=\frac{n_{2k}(t+1)}{n_2(t)}$
.	.	.	.	..	.
.	.	.	.	..	.
.	.	.	.	..	.
k	$n_k(t)$	$n_{k1}(t+1)$ $p_{k1}=\frac{n_{k1}(t+1)}{n_k(t)}$	$n_{k2}(t+1)$ $p_{k2}=\frac{n_{k2}(t+1)}{n_k(t)}$	...	$n_{kk}(t+1)$ $p_{kk}=\frac{n_{kk}(t+1)}{n_k(t)}$

Table 3. Transition probabilities matrix (P) with transition probabilities p_{ij} (from: [67])

The "state space" and time can be either continuous or discrete [62, 63]. A Markov process is called stationary or homogeneous if $P(t)=p_{ij}$, $\forall\ t \in \mathbb{N}$, otherwise it is called non-homogeneous. A non-homogeneous Markov process undergoes cyclic behaviour if there exists $d \in \mathbb{N}$ such that $P(ad+s)=P(s)$, $\forall\ t \in \mathbb{N}$ and s={1,2,...,d-1} [64]. For ecosystems with seasonal cycles d=4, let P(0), P(1), P(2) and P(3) be the transition matrices from winter to spring, spring to summer, summer to autumn and autumn to winter, respectively. Then, the transition matrices from winter to winter (P_0), spring to spring (P_1), summer to summer (P_2) and autumn to autumn (P_3) are given by:

$$P_0 = P(0)\, P(1)\, P(2)\, P(3)$$
$$P_1 = P(1)\, P(2)\, P(3)\, P(0)$$
$$P_2 = P(2)\, P(3)\, P(0)\, P(1)$$
$$P_3 = P(3)\, P(0)\, P(1)\, P(2)$$

The matrices P_i, i={0,1,2,3} are stochastic and converge if they are regular, and the convergence is geometrically fast. Their limits (P_i^t) are independent of the starting season, meaning that lim(winter to winter)=lim(spring to winter)=.. =lim(autumn to winter), and give the probabilities of the system in "statistical equilibrium" for each season [65-67].

2.4.2. *Case study*

A study of zoobenthos dynamics from the Thermaikos Gulf, Greece, is presented in this section [37, 67]. In this study, the following aspects are examined: the seasonal variation of zoobenthos richness (number of species/0.2 m^2), abundance (individuals/0.2 m^2) and diversity. Concerning zoobenthos richness, four states were chosen: {A_1(up to 10 species), A_2 (from 11 to 15 species), A_3 (from 16 to 20 species) and A_4 (more than 20 species). The definition of the "state space" was based on the literature. For example, fewer than 10 species have been reported in pollutant waters with oxygen concentrations less than 1.4 ml/l [68, 69], while in gulfs similar to the Thermaikos Gulf, without pollution effects, more than 20 species have been reported [70]. The definition of A_3 and A_4 was based on data from areas under eutrophication processes, where the number of species varied seasonally between 10 and 20 [70-72]. On abundance, four states were defined {(B_1 (less than 90 individuals), B_2 (from 91-140), B_3 (from 141 to 200) and B_4 (more than 200)}, following the same assumptions as before when considering diversity. The estimated matrices with transition probabilities from each season to the next, starting from winter to spring, are presented in Table 4.

	From winter to spring				From spring to summer			
	A_1	A_2	A_3	A_4	A_1	A_2	A_3	A_4
A_1	0.5	0.5	0	0	0.584	0.25	0.083	0.083
A_2	0.714	0.143	0.143	0	0.112	0.444	0.333	0.111
A_3	0.461	0.461	0.078	0	0	0.333	0.667	0
A_4	0	0.5	0.25	0.25	0	0	0.5	0.5
	From summer to autumn				From autumn to winter			
	A_1	A_2	A_3	A_4	A_1	A_2	A_3	A_4
A_1	0.5	0.5	0	0	0.584	0.25	0.083	0.083
A_2	0.714	0.143	0.143	0	0.112	0.444	0.333	0.111
A_3	0.461	0.461	0.078	0	0	0.333	0.667	0
A_4	0	0.5	0.25	0.25	0	0	0.5	0.5

(a)

	From winter to spring				From spring to summer			
	B_1	B_2	B_3	B_4	B_1	B_2	B_3	B_4
B_1	0.833	0.167	0	0	0.786	0.071	0.143	0
B_2	0.712	0.224	0.064	0	0.4	0.2	0.2	0.2
B_3	0.222	0.222	0.334	0.222	0.2	0.4	0.2	0.2
B_4	0.571	0.143	0.286	0	0	0	0.5	0.5
	From summer to autumn				From autumn to winter			
	B_1	B_2	B_3	B_4	B_1	B_2	B_3	B_4
B_1	0.571	0.143	0.214	0.072	0.333	0.167	0.25	0.25
B_2	0.4	0.2	0.2	0.2	0.25	0.25	0.25	0.25
B_3	0.25	0.25	0.25	0.25	0	0.4	0.4	0.2
B_4	0	0	0.333	0.667	0	0	0.6	0.4

(b)

Table 4. Transition probabilities for species richness (a) and abundance (b)

The matrices P_i, i={0, 1, 2, 3} and their limits (P_i^t) were estimated (Fig. 11). Additionally, in order to test if the number of individuals (abundance) and the number of species (richness) are independent, correlation coefficients were estimated and, furthermore, simple regression models were applied to each set of data separately. In all cases, the statistical tests confirm that the two variables are independent ($p<0.05$). So, the joint probability P(A_i and B_i) in statistical equilibrium could be estimated as follows:

$$P\left(A_i \text{and } B_i\right) = P\left(A_i\right) * P\left(B_i\right) \rightarrow i = \{1,2,3,4\}.$$

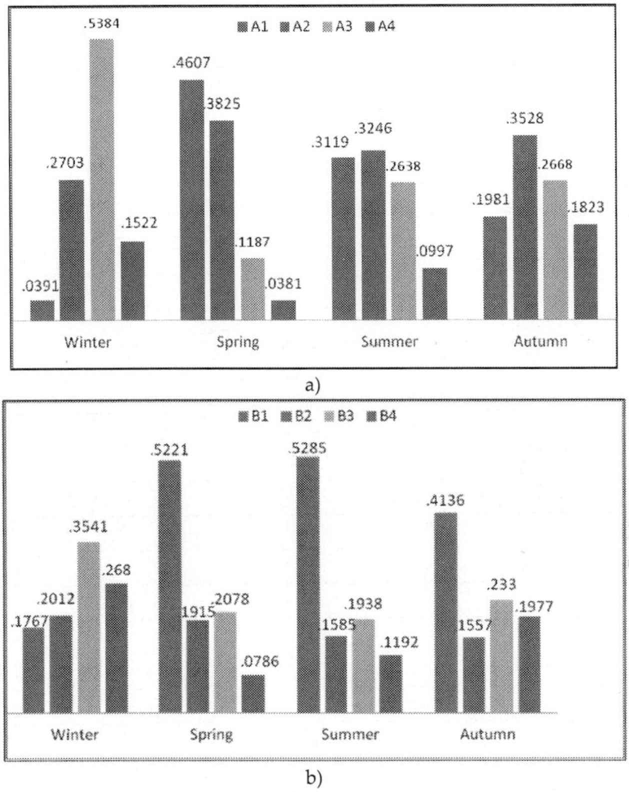

Figure 11. The limits of transition probabilities' matrices for each season: a) for species richness and b) for abundance (population density). The values of the probabilities are represented in the y axes.

According to Fig. 11, it could be said that winter is the best season, because the probabilities of the system being in the more "rich states" (A_3, A_4 and B_3 and B_4) are 0.6916 and 0.6221, respectively. At the same time, the probability of the system to have more than 16 species and more than 141 individuals {(A_3 and B_3) or (A_3 and B_4) or (A_4 and B_3) or (A_4 and B_4)}, in steady state, is higher than 0.42 (Fig. 12). On the other hand, during spring, the probabilities of A_1 (=0.46) and B_1 (=0.5221) taking their highest values among all the seasons (Fig. 11) and, furthermore, the probability of the system having fewer than 15 species and simultaneously less than 90 individuals, is higher than 0.6 (Fig. 12). In relation to these values, it could be suggested that in spring, strong environmental influences tend to reduce both the species richness and the populations density. During summer, it can be observed (Table 3) that transition probabilities to A_1 and A_2 still have high values, but by contrast transition probabilities to A_3 and A_4 are higher than the respective ones in spring. Concerning abundance, it could be observed (Fig. 11) that B_1 is dominant, with the highest probability value (=0.5285). The probability of the system having fewer than 15 species and fewer than 90 individuals is 0.44. So, during summer, population density seems to stabilize in low densities, similar to the behaviour observed in the spring. During autumn, the "richest state", A_4, takes its highest probability value (=0.1823) and simultaneously A_3 and A_4 are higher than those of spring and summer. In addition, B_1 probability is lower than that observed for summer, although it continues to dominate (Fig. 11). The joint probability of the system having 11-15 species and up to 140 individuals is estimated at 0.353 (Fig. 12). So, it could be considered that a tendency towards recovery starts from autumn and continues to winter.

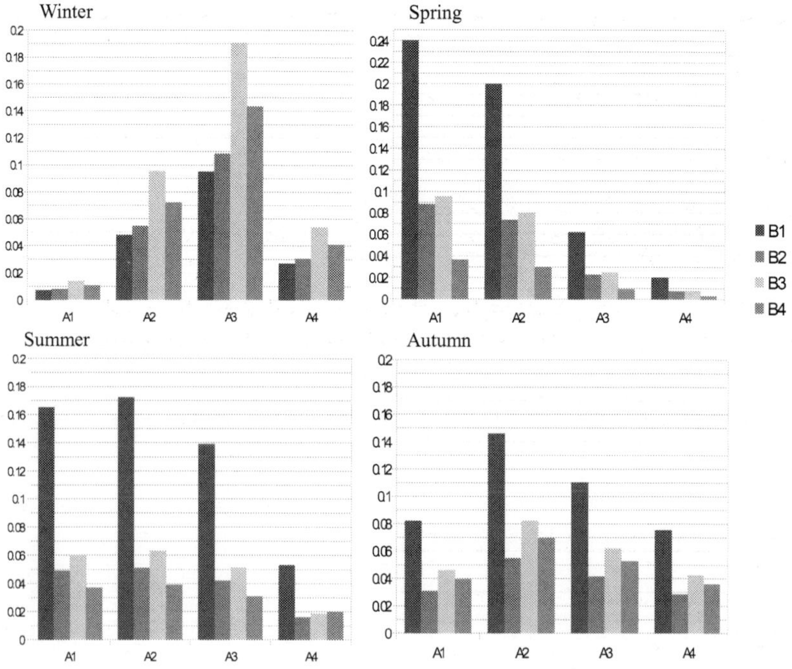

Figure 12. Joint probabilities A_i and B_i i={1,2,3,4}, for each season separately. The values of the probabilities are represented in the y axes.

The above discussion points out that the composition of the benthic community of zoobenthos from the Thermaikos Gulf may be regulated by seasonal sequences of extinction and recolonization. The two "forcing variables" in the Thermaikos Gulf are oxygen concentration and temperature. So, species with tolerance to high temperature, low oxygen and high pollutant concentrations may regulate their births from spring to summer, while species with no tolerance to low oxygen or/and high temperatures may regulate their births from autumn to winter, when the waters are saturated with oxygen. The low values of abundance (population density) were observed during spring to autumn, and may be caused by mass mortalities due to oxygen depletion in the bottom waters [67].

3. Conclusions

The fact that the TRIX index has been applied in many areas suggests its use as a tool not only for the classification of marine waters but also for the comparison between data from different regions. The problem of its calibration arises when marine ecological processes of the study region differ from those of the Northern Adriatic Sea. In such cases, a re-scaled index like the KALTRIX seems to be more sensitive than TRIX to minor changes in the parameters, but as the index was re-scaled its results are not directly comparable with those of other regions. UNTRIX is also a sensitive index, but the problem of the reference site and its characteristics is evident. So, it could be said that if the interest of a manager is to compare a marine ecosystem before and after a strong human influence (like WWTP), then more sensitive indices like KALTRIX or UNTRIX should be used; otherwise, the use of the TRIX index gives the advantage of enabling direct comparisons between different marine ecosystems.

An alternative approach for marine water classification and management is the use of piecewise linear regression models with breakpoints. Such models are able to express discontinuities in ecosystems and to detect critical thresholds in coastal ecosystems.

Concerning the case of the numerical dynamic hydrobiological model, the dispersion of harmful cells after a HAB episode in a coastal basin was investigated. As expected, the wind speed and direction significantly affect the dispersion of an algal bloom. According to the results of the present work, the location of an initial outbreak in combination with the topography of a coastal basin play an important role in the appearance and the dispersion of an algal bloom, because winds with the same speed induce two different patterns of dispersion and different phytoplankton abundances (or concentrations) in space and time. The application of the dynamic hydrobiological model simulating a real episode showed that after a phytoplankton bloom at the north-east coasts of the Thermaikos Gulf (with variable wind forcing over the coastal basin and prevailing northern winds), concentrations of harmful algal cells reached (a) the north-west and west coasts of the gulf where the largest part of the mussel culture of Greece lies, and (b) the west coasts of the gulf with important bathing beaches, within a period of 10 days. For 20 days after the bloom, the mass of *Dinophysis* (HAB) dispersed in quite large areas of the outer Thermaikos Gulf. Concerning the application of the above model to the larger area of the extended Thermaikos Gulf, the model runs showed that, after an algal bloom, the stronger the winds are, the lower the concentration of the dispersed phytoplanktonic cells is expected to be. This means that stronger winds more effectively disperse the masses, "diluting" the large hazardous populations. Moreover, when the source is located in the inner part of the gulf, the dispersion of the bloom does not seem to reach to the outer parts of the basin under north or south winds. From the applications of the dynamic hydrobiological models described here, it can be said that the spreading of phytoplankton cells in the large area of a gulf or a coastal basin can be investigated analytically, so that the essential measures of prevention can be applied for each different case.

Finally, if time-dependent uncertainty and randomness affect the processes under investigation, and seasonal cycles are displayed, then non-homogeneous Markovian models with cyclic behaviour may be useful tools for their description. Furthermore, these type of models may be used as an alternative approach to the study of diversity and stability of such marine ecosystems.

Acknowledgements

We owe many thanks to the Forestry Research Institute for providing us with wind data for the period of January 2000. We also wish to thank Dr O. Gotsis-Skretas, Dr Ch. Kontoyiannis and Dr A Pavlidou from the Hellenic Center for Marine Research for providing the data from S. Evoikos Gulf.

Author details

Dimitrios P. Patoucheas[1*] and Yiannis G. Savvidis[2]

*Address all correspondence to: dpatouh@bio.auth.gr

1 Ministry of Education and Religious Affairs, Secondary Education of Thessaloniki, Thessaloniki, Greece

2 Department of Civil Engineering of Technological Education, Alexander Technological Educational Institute of Thessaloniki, Thessaloniki, Greece

References

[1] Xie H., Lazure P., Gentien P. Small scale retentive structures and Dinophysis. Journal of Marine Systems 2007: 64(1-4), 173-188.

[2] Ganoulis J. G. Water quality assessment and protection measures of a semi-enclosed coastal area: the Bay of Thermaikos (NE Mediterranean Sea). Mar. Poll. Bull. 1991: 23, 83-87.

[3] Ganoulis J. G. Engineering Risk Analysis of Water Pollution. VCH Verlagsgesellschaft mbH, Weinheim, Germany, 1994.

[4] Ryan J. P., Fischer A. M., Kudela R. M., Gower J. F. R., King S. A., Marin R. III, Chavez F. P. Influences of upwelling and downwelling winds on red tide bloom dynamics in Monterey Bay, California. Continental Shelf Research 2009: 29, 785–795.

[5] Patoucheas D. P., Savvidis Υ. G. A hydrobiological model as a tool for the detection of HAB episodes. Application to Thermaikos Gulf. Fresenius Environmental Bulletin 2010: 19(9b), 2100-2108.

[6] Savvidis Υ. G., Patoucheas D. P., Nikolaidis G., Koutitas C. G. Modeling the dispersion of harmful algal bloom in the Thermaikos Gulf (NW Aegean Sea). Global NEST Journal (Global Network of Environmental Science and Technology) 2011: 13(2), 119-129.

[7] Vollenweider R. A., Giovanardi F., Montanari G., Rinaldi A. Characterization of the trophic conditions of marine coastal waters, with special reference to the NW Adriatic Sea: Proposal for a trophic scale, turbidity and generalized water quality index. Environmetrics 1998: 9(3), 329-357.

[8] Penna N., Capellacci S., Ricci F. The influence of the Po River discharge on phytoplankton bloom dynamics along the coastline of Pesaro (Italy) in the Adriatic Sea. Marine Pollution Bulletin 2004: 48(3), 321-326.

[9] Giovanardi F., Vollenweider R. A. Trophic conditions of marine coastal waters: Experience in applying the Trophic Index TRIX to two areas of the Adriatic and Tyrrhenian Seas. Journal of Limnology 2004: 63(2), 199-218.

[10] Pettine M., Casentini B., Fazi S., Giovanardi F., Pagnotta R. A revisitation of TRIX for trophic status assessment in the light of the European Water Framework Directive: Application to Italian coastal waters. Marine Pollution Bulletin 2007: 54(9), 1413-1426.

[11] Nikolaidis G., Moschandreou K. Patoucheas, D. P. Application of a trophic index (TRIX) for water quality assessment at Kalamitsi Coasts (Ionian Sea) after the operation of the wastewater treatment plant. Fresenius Environmental Bulletin 2008: 17(11 B), 1938-1944.

[12] Balkis N., Toklu-Alici B., Balci M. Evaluation of ecological quality status with the trophic index (TRIX), values in the coastal waters of the Gulfs of Erdek and Bandirma in the Marmara Sea. Ecological Water Quality Water Treatment and Reuse 2012: 1-22.

[13] Vascetta M., Kauppila P., Furman E. Aggregate indicators in coastal policy making: Potentials of the trophic index TRIX for sustainable considerations of eutrophication. Sustainable Development 2008: 16(4), 282-289.

[14] Moncheva S., Gotsis-Skretas O., Pagou K., Krastev A. Phytoplankton blooms in Black Sea and Mediterranean coastal ecosystems subjected to anthropogenic eutrophication: Similarities and differences. Estuarine, Coastal and Shelf Science 2001: 53(3), 281-295.

[15] Herrera-Silveira J. A., Morales-Ojeda S. M. Evaluation of the health status of a coastal ecosystem in southeast Mexico: Assessment of water quality, phytoplankton and submerged aquatic vegetation. Marine Pollution Bulletin 2009: 59(1), 72-86.

[16] Taebi S., Etemad-Shahidi A., Faradi G. A. Examination of three eutrophication indices to characterize water quality in the north east of Persian Gulf. J. Coastal Res. 2005: 42, 405-411.

[17] Zoriasatein N., Jalili S., Poor F. Evaluation of ecological quality status with the trophic index (TRIX) values in coastal area of Arvand, northeastern of Persian Gulf, Iran. World 2013: 5(3), 257-262.

[18] Primpas I., Karydis M. Scaling the trophic index (TRIX) in oligotrophic marine environments. Environmental Monitoring and Assessment 2011: 178(1-4), 257-269.

[19] Toms J. D., Lesperance M. L. Piecewise regression: A tool for identifying ecological thresholds. Ecology 2003: 84(8), 2034-2041.

[20] Nikolaidis G., Patoucheas D. P., Moschandreou K. Estimating breakpoints of Chl-a in relation with nutrients from Thermaikos Gulf (Greece) using piecewise linear regression. Fresenius Environmental Bulletin 2006: 15(9 B), 1189-1192.

[21] Gaston A. J., Smith P. A., Provencher J. F. Discontinuous change in ice covers in Hudson Bay in the 1990s and some consequences for marine birds and their prey. ICES Journal of Marine Science: Journal du Conseil 2012: fss 040.

[22] Prasad A. K., Chai L., Singh R. P., Kafatos M. Crop yield estimation model for Iowa using remote sensing and surface parameters. International Journal of Applied Earth Observation and Geoinformation 2006: 8(1), 26-33.

[23] Cavanaugh K. C., Kellner J. R., Forde A. J., Gruner D. S., Parker J. D., Rodriguez W., Feller I. C. Poleward expansion of mangroves is a threshold response to decreased frequency of extreme cold events. Proceedings of the National Academy of Sciences of the USA 2014: 111(2), 723-727.

[24] Morley S. A., Lai C. H., Clarke A., Tan K. S., Thorne M. A., Peck L. S. Limpet feeding rate and the consistency of physiological response to temperature. Journal of Comparative Physiology B 2014: 1-8.

[25] Sany S. B. T., Hashim R., Rezayi M., Salleh A., Safari O. A review of strategies to monitor water and sediment quality for a sustainability assessment of marine environment. Environmental Science and Pollution Research 2014: 21(2), 813-833.

[26] Gurel M., Ekdal A., Erturk A., Tanik A. Efforts towards setting eutrophication assessment criteria for coastal marine ecosystems. International Journal of Environment and Pollution 2005: 23(3), 325-335.

[27] Koukaras K., Nikolaidis G. Dinophysis blooms in Greek coastal waters (Thermaikos Gulf, NW Aegean Sea). Journal of Plankton Research 2004: 26(4), 445-457.

[28] Sah G., Mohanty, A. K., Samantara M. K., Satpathy K. K. Seasonality in the distribution of dinoflagellates with special reference to harmful algal species in tropical coastal environment, Bay of Bengal. Environmental Monitoring and Assessment 2014: 1-18.

[29] Koutitas C. Mathematical models in Coastal Engineering. Pentech Press Limited, London (UK), 1988.

[30] Savvidis Y. G, Krestenitis Y. N., Koutitas C. G. Modeling the water mass exchange through navigational channels connecting adjacent coastal basins. Application to the Channel of Potidea (North Aegean Sea). Annales Geophysicae 2005: 23, 231-238.

[31] Smagorinsky J. General circulation experiments with the primitive equations. I. The basic experiment. Monthly Weather Review 1963: 91, 99-164.

[32] Jeng S. W. Two-dimensional Random Walk Model for Pollutant Transport in Natural Rivers. PhD thesis. University of Texas, 1986.

[33] Dimou K. N., Adams E. E. A random-walk, particle tracking model for well-mixed estuaries and coastal waters. Estuarine Coastal and Shelf Science 1993: 33, 99-110.

[34] Savvidis Y., Koutitas C. Simulation of transport & fate of suspended matter along the coast of Agathoupolis (N. Greece). Proceedings of the 5th International Conference: Protection and Restoration of the Environment, Thasos, Greece, 2000.

[35] Savvidis Y., Koutitas C., Krestenitis Y. N. Development and application of a three dimensional cohesive sediment transport mathematical model. Journal of Marine Environmental Engineering 2001: 6, 229-255.

[36] Krestenitis Y. N., Kombiadou K. D., Savvidis Y. G. Modelling the cohesive sediment transport in the marine environment: The case of Thermaikos Gulf. Ocean Science 2007: 3, 91-104.

[37] Patoucheas D. P. Ecological model of Thermaikos Gulf. PhD thesis. Aristotle University of Thessaloniki, Department of biology, Thessaloniki, Greece, 1995

[38] Patoucheas D. P., Dasiou V. Spatial and temporal growth rate changes of Thermaikos Gulf dinoflagellates: Simulation and analysis. 1st Scientific Conference of EFMS: Oceanographical Aspects for a Sustainable Mediterranean, Athens, September 27-29th, 2002.

[39] Eppley R.W. Temperature and phytoplankton growth in the sea. Fish. Bull. 1972: 70, 1063-1085.

[40] Bissinger J. E., Montagnes D. J. S., Sharples J., Atkinson D. Predicting marine phytoplankton maximum growth rates from temperature: Improving on the Eppley curve using quantile regression. Limnology and Oceanography 2008: 53(2), 487-493.

[41] Steele J. H. Environmental control of photosynthesis in the sea. Limn. Oceanogr. 1962: 7, 137-150.

[42] Anonymous. Observations meteorologiques de Thessaloniki 2000-2001. Ann. de'l Inst. Meteir. et Climatol. 2001: 69.

[43] Putt M., Rivkin R., Prezelin B. B. Effects of altered photic regimes on diel patterns of species-specific photosynthesis - I. Comparison of polar and temperate phytoplankton. Marine Biology 1988: 97, 435-443.

[44] Andersen P., Nival P. Modelling of phytoplankton population dynamics in an enclosed water column. J. Mar. Biol. Ass. UK 1989: 69, 625-646.

[45] Landry M. R., Hassett R. P. Estimating the grazing impact of marine micro-zooplankton. Marine Biology 1982: 67, 283-288.

[46] Landry M. R., Peterson W. K., Lorenzent C. J. Zooplankton grazing, phytoplankton growth, and export flux: Inferences from chlorophyll tracer methods. ICES J. Mar. Sci. 1995: 52, 337-345.

[47] Peperzak L., Duin R. N. M., Colijn F., Gieskes W. W. C. Growth and mortality of flagellates and non-flagellate cells of *Phaeocystis globosa* (Prymnesiophyceae). Journal of Plankton Research 2000: 22(1), 107-120.

[48] Garcés E., Vila M., Masó M., Sampedro N., Giacobbe M. G., Penna A. Taxon-specific analysis of growth and mortality rates of harmful dinoflagellates during bloom conditions. Marine Ecology Progress Series 2005: 301, 67-79.

[49] Huret M., Gohin F., Delmas D., Lunven M., Garçon V. G. Use of SeaWiFS data for light availability and parameter estimation of a phytoplankton production model of the Bay of Biscay. Journal of Marine Systems 2007: 65, 509–531.

[50] Eppley R. W., Rogers J. N., McCarthy J. J. Half saturation constants for uptake of nitrate and ammonium by various phytoplankton. Limnol. Oceanogr. 1969: 14, 912-920.

[51] Pasciak W. J., Gavis J. Transport limitation of nutrient uptake in phytoplankton. Limnol. Oceanogr. 1974: 19, 881-885.

[52] Taft J. L., Taylor W. R., McCarthy J. J. Uptake and release of phosphorus by phytoplankton in the Chesapeake Bay estuary, USA. Marine Biology 1975: 33, 21-32.

[53] Yamamoto T., Tarutani K. Growth and phosphate uptake kinetics of the toxic dinoflagellate *Alexandrium tamarense* from Hiroshima Bay in the Seto Inland Sea, Japan. Phycological Research 1999: 47(1), 27-32.

[54] Yamamoto T., Oh S. J., Kataoka Y. Growth and uptake kinetics for nitrate, ammonium and phosphate by the toxic dinoflagellate *Gymnodinium catenatum* isolated from Hiroshima Bay, Japan. Fisheries Science 2004: 70(1), 108–115.

[55] Huthnance J. M. et al., editors. PROFILE - Processes in Regions of Freshwater Influence Final Report. POL Internal Document No 102 - Thermaikos Bay, 1997.

[56] Hyder P. J., Simpson H., Christopoulo S., Krestenitis, Y. The seasonal cycles of stratification in the Thermaikos Gulf Region of Freshwater Influence (ROFI), North-west Aegean. Continental Shelf Research 2002: 22, 2573-2597.

[57] Gisselson L. Å., Carlsson P., Granéli E., Pallon J. Dinophysis blooms in the deep euphotic zone of the Baltic Sea: Do they grow in the dark? Harmful Algal 2002: 1(4), 401-418.

[58] Patten B.C. Energy cycling in the ecosystem. Ecological Modelling 1985: 28(1), 1-71.

[59] Hill M. F., Witman J. D., Caswell H. Markov chain analysis of succession in a rocky subtidal community. The American Naturalist 2004: 164(2), 46-61.

[60] Pearse W. D., Purvis A., Roy D. B., Stamatakis A. Modelling ecological communities as if they were DNA. arXiv preprint arXiv 2014: 1403.7668.

[61] Kershenbaum A., Blank L., Sinai I., Merilä J., Blaustein L., Templeton A. R. Landscape influences on dispersal behaviour: A theoretical model and empirical test using the fire salamander, *Salamandra infraimmaculata*. Oecologia 2014: 175(2), 509-520.

[62] Maaita J. O., Tsaklidis G. Interpretation of the continuous time homogeneous Markov system with fixed size as an elastic solid. 6th European Solid Mechanics Conference ESMC 2006, Budapest, Hungary, 2006.

[63] Vasiliadis G., Tsaklidis G. On the distributions of the state sizes of discrete time homogeneous Markov systems. Methodology and Computing in Applied Probability 2008: 10(1), 55-71.

[64] Vassiliou P. C. Cyclic behaviour and asymptotic stability of non-homogeneous Markov systems. Journal of Applied Probability 1984: 315-325.

[65] Vassiliou P. C. Asymptotic variability of nonhomogeneous Markov systems under cyclic behaviour. European Journal of Operational Research 1986: 27(2), 215-228.

[66] Vassiliou P. C., Tsaklidis G. The rate of convergence of the vector of variances and covariances in non-homogeneous Markov systems. Journal of Applied Probability 1989: 776-783.

[67] Patoucheas D.P., Stamou G. Non homogeneous Markovian models in ecological modelling: A study of zoobenthos dynamics in Thermaikos Gulf, Greece. Ecological Modelling 1993: 66(3), 197-215.

[68] Rosenberg R. Benthic macrofaunal dynamics, production, and dispersion in an oxygen-deficient estuary of west Sweden. Journal of Experimental Marine Biology and Ecology 1977: 26(2), 107-133.

[69] Zarkanellas A. J. The effects of pollution-induced oxygen deficiency on the benthos in Elefsis Bay, Greece. Marine Environmental Research, 1979: 2(3), 191-207.

[70] Bogdanos C., Satsmadjis J. The macrozoobenthos of an Aegean embayment. Thalassographica, 1983: 6, 77-105.

[71] Lie U., Kisker D. S. Species composition and structure of benthic infauna communities off the coast of Washington. Journal of the Fisheries Board of Canada 1970: 27(12), 2273-2285.

[72] Diapoulis A., Bogdanos C. Preliminary study of soft substrate macrozoobenthos and marine flora in the Bay of Gera (Lesvos Island, Greece). Thalassographica, 1983: 6, 127-139.

Chapter 3

Alternative Uses of Cavitating Jets

José Gilberto Dalfré Filho, Maiara Pereira Assis and
Ana Inés Borri Genovez

Additional information is available at the end of the chapter

http://dx.doi.org/10.5772/59136

1. Introduction

Cavitation is a phase transformation occurring in a fluid system under certain conditions. If the dynamic alteration of the absolute static pressure reaches or drops below the vapour pressure of the liquid, vapour bubbles are formed inside the liquid and can collapse as they move to a high-pressure region [1]. When the vapour bubbles collapse, shock waves are produced that propagate at the speed of sound through the liquid [2]. So, cavitation can produce undesirable effects such as noise, vibration, pressure fluctuation, erosion and efficiency loss in a hydraulic system.

The pressure difference due to the dynamic effect of the fluid motion is proportional to the square of the relative velocity. This can be written in the form of the usual pressure coefficient (C_p), presented in the Equation 1 [3]:

$$C_p = \frac{(p - p_o)}{\rho v_o^2 / 2} \tag{1}$$

where ϱ is the density of the liquid; v_o and p_o are the velocity and pressure of an undisturbed liquid, respectively, and (p-p_o) is the pressure differential due to dynamic effects of fluid motion.

If pressure p reaches a minimum, $p=p_{min}$ (see Figure 1), the pressure coefficient will be minimum, and a set of conditions can be created so that p_{min} drops to a value where cavitation can begin. This can be accomplished by raising the relative velocity v_o for a fixed value of the pressure p_o or lowering p_o with v_o remaining constant. If surface tension is ignored, the pressure p_{min} will be the pressure inside the cavity. This pressure will be the bubble pressure.

INTECH
open science | open minds

If we consider that cavitation will occur when the normal stresses at a point in the liquid are reduced to zero, then the bubble pressure will assume the value of the vapour pressure, p_v. Then, the cavitation index (σ) is the theoretical value of the negative pressure coefficient, Equation 2:

$$\sigma = \frac{p_o - p_v}{\rho v_o^2 / 2} = -(C_p)_{\min} \tag{2}$$

The cavitation index can designate the probability of a system to cavitate and establish different intensity levels of cavitation in this system.

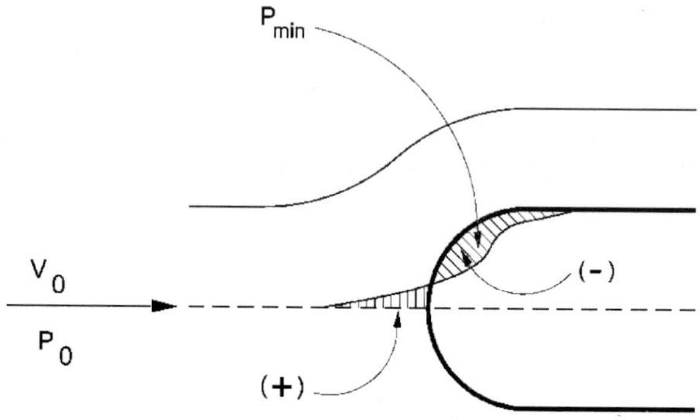

Figure 1. Pressure distribution in a submerged body [4].

As the pressure is increased, the bubble diameter decreases from the original size reaching a minimum size. The process takes place until the bubble diameter becomes microscopic. If the bubble collapses, then shock waves form with celerity equal to the speed of sound in water. If a boundary is close to where the bubble imploded, it will deform into a microjet. The velocity of this microjet is high and the shockwave produces high pressure responsible for the cavitation damage of a surface [5]. The collapse of the cavities formed by the cavitating jet generates high-pressure waves, estimated to be approximately 69.00 MPa [6] and high-speed microjets (above 100 m/s), all of which have a significant amount of destructive power [7].

Flow conditions leading to the onset of cavitation are generally conservative in predicting damage. The severity of the damage may be related to both intensity of cavitation and exposure time [8].

In hydraulic structures, as the high velocity flow passes over the many irregularities that exist in the concrete surface, cavitation can commence and consequently, damages may occur. Therefore, the material properties of the surface have to be improved to provide adequate resistance either during the construction phase or when the substitution of the eroded concrete is required.

Particularly of interest, in developing countries, hydropower will continue to play a significant role in supplying energy. Therefore, as energy demands are continuously increasing, new hydropower plants (dams and appurtenances) are being constructed. Nevertheless, the operation and maintenance of these structures, which include the spillways and tunnels under high velocity flows, are a great issue. Therefore, the material properties have to be improved to provide adequate resistance, either during the construction phase or substitution phase. For example, the substitution of a great area of eroded concrete was required in the case of Porto Colombia hydropower spillway and dissipation basin in the Grande River, Brazil [7].

An additional complexity is observed in hydraulic structures due to the simultaneous effect of cavitation and high impact of the flow [9,10,4]. Evaluation of the concrete resistance erosion in hydraulic structures is essential to guarantee adequate operation. It was suggested by [11] defining a methodology to appropriately test the materials, when submitted to cavitation, to be used in hydraulic structures.

In this chapter the authors discuss the use of high velocity cavitating jets to determine the erosion in high performance concretes for hydraulic structures. Moreover, two alternatives uses for cavitating jets are presented: 1) the inactivation of *Escherichia coli,* and, 2) the decomposition of persistent compounds in water.

2. Use of the cavitating jet to test erosion in surfaces

2.1. Introduction

Evaluation of the concrete erosion resistance in hydraulic structures is essential to guarantee adequate operation, and tests should be performed under the same operation conditions. Various techniques can be used to induce cavitation, such as, ultrasonic methods, hydrodynamic methods, and high-speed/high-pressure homogenization. In the hydrodynamic cavitation, pressure variations are produced using the geometry of the system, while in the acoustic cavitation pressure variations are effected using sound waves.

Concrete resistance to cavitation damage was investigated by [12-14] using a Venturi device. A chamber was used by [15] to test concrete samples subjected to short-duration cavitation. An alternative test for erosion evaluation uses water cavitating jet technology [16,17] to achieve short test time. A flow cavitation chamber was used by [18-20] to test concrete and different rock samples subjected to short-duration cavitation. Water cavitating jet technology was firstly used to clean surfaces. For example, [21] compared the efficiency of conventional non-cavitating jets to cavitating jets.

Regardless several researchers have worked with the cavitation jet apparatus, to test concretes submitted to the erosive effect of cavitation, the appropriate nozzle specifications and sample dimensions must be experimentally determined.

2.2. The cavitating jet apparatus

The cavitating jet apparatus uses a nozzle specially designed to produce cavitation, combining high-velocity flows and cavitation (with an appropriate cavitation index). Because the collapse of bubbles is concentrated over a microscopic area, localized stresses are produced, providing the cavitating jet with a great advantage over steady non-cavitating jet operating at the same pressures and flow rates [22]. Figure 2 shows a schematic of the formation of a cavitating jet. As the high velocity flow leaves the orifice nozzle, eddies are formed between the high velocity layer of the jet and the surrounding liquid. If a nucleus in the water is captured in one of these eddies and the pressure inside the eddy drops to vapour pressure, then the nucleus will begin to grow. However, if the pressure remains near the vapour pressure long enough for the nuclei to reach the critical diameter, then it begins to grow almost by vaporization. As long as the size of the vapour cavity increases, the strength of the eddy is destroyed and the rotational velocity decreases. Then, the surrounding bubble pressure is no longer the vapour pressure. Inside the cavity the pressure remains at vapour pressure and cavity surroundings at hydrostatic pressure. Consequently, the cavity becomes unstable and it collapses inward.

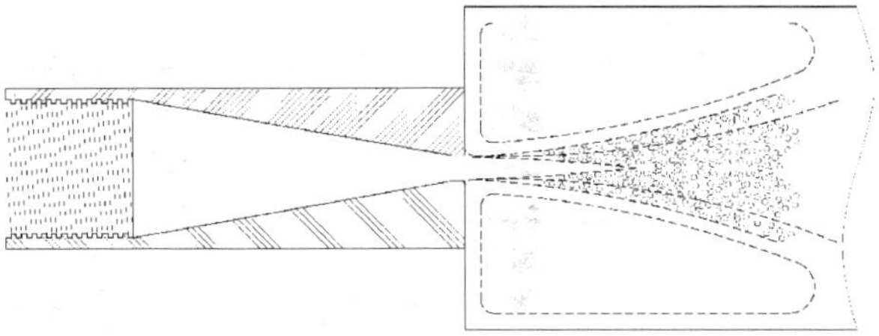

Figure 2. Schematic flow of a submerged jet [6].

Now, consider in Figure 3 a spherical vapour bubble of initial radius b and internal pressure P_o. Later on, the bubble acquires the radius r_e at pressure P_e. If a small amount of air is in the liquid, such that the cavities are filled exclusively by vapour, the bubble growth and collapse is intense and causes severe damages to the vicinity [6].

The pressure magnitude generated at the solid boundary is expected to be a main factor in estimating the erosion efficiency of a cavitating jet impinging on a solid boundary as shown in Figure 3.

For a steady water jet (v_j),the generated pressure magnitude (P_s) can be estimated from the stagnation pressure, as in Equation 3:

$$P_s = \frac{1}{2} \rho v_j^2 \tag{3}$$

A moderate level pressure would be inappropriate for testing erosion rates. Depending on the standoff distance, water droplets formation can occur and the impact pressure can be larger

http://dx.doi.org/10.5772/59136

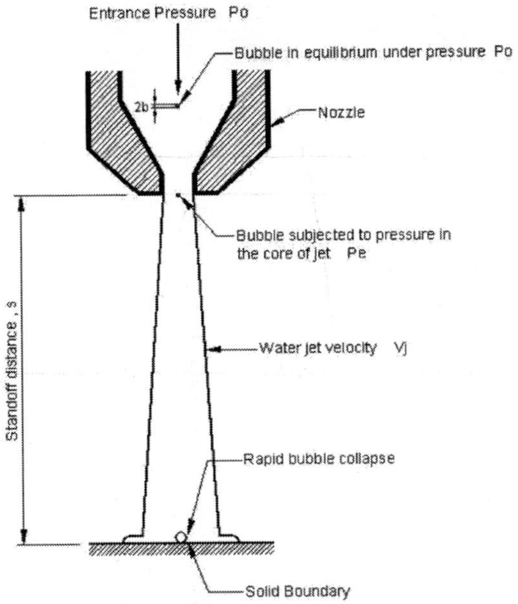

Figure 3. Cavitating jet formation and bubble collapse (Adapted from [23])

than P_s. It was indicated that the transient pressure (P_a) between the cavity implosion and the solid boundary could be approximated by the "water hammer" equation, as in Equation 4:

$$P_a = \rho c v_j \tag{4}$$

with c as the sound velocity in water.

However, the pressure magnitude generated with the impact of a cavitating jet (P_b) is different from that generated by the impact of distinct droplets. According to [21] the maximum pressure P_b develops if a bubble collapses in an incompressible flow, under isothermal conditions, resulting in equation (5):

$$P_b = \frac{P_S}{6,35} e^{\left(1/\alpha\right)} \tag{5}$$

where α relates the gas pressure inside the bubble in the beginning of collapse under the pressure P_s. Therefore, the cavitating jet erosion efficiency R* is related to P_b and P_a in Equation (6):

$$R^* = \frac{P_b}{P_a} \tag{6}$$

Equation (5), however, does not take into account the geometry and finishing of the nozzle. It was verified by [10] that the finish and the geometry of the nozzle influence the erosion performance of the equipment. The author accomplished a series of tests with a cavitating jet apparatus, verifying the influence of these two variables in aluminium and concrete samples. Therefore, in Equation (7) a geometric efficiency factor η was introduced in the cavitation erosion efficiency [7] to calculate the total efficiency η :

$$\varepsilon = \eta R^* \tag{7}$$

2.3. Experimental setup

To simulate the cavitation phenomenon and to evaluate the erosion in samples, a cavitating jet apparatus was constructed [10] in the Laboratory of Hydraulics and Fluid Mechanics at the State University Campinas, Brazil. Figure 5 and 6 shows a schematic representation and a photo of the test facility. A high-pressure displacement pump (kept at pressure 12.00 MPa) conducts water to the facility and to the metallic chamber provided with two windows for visualization of the tests. A pressure-regulating valve guaranteed safe operation. At the end of the high-pressure pipe and inside of the inactivation chamber, the nozzle with a hole (diameter of 2.00 x 10^{-3} m) was positioned at the end of the high-pressure pipe.

The nozzle specifications must be established according to its use and they are determined experimentally. The researcher [10] used three nozzle geometries (20° conical, 132° conical and circular) with both shaped and rounded edges (Figure 4).

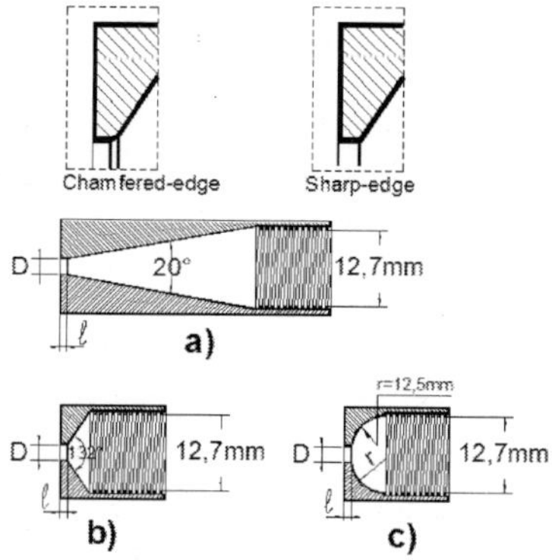

Figure 4. Nozzle geometries (a) conical 20°, (b) conical 132°, (c) circular [7].

The facility is supplied with no re-circulated water. The samples are placed in the chamber subjected to the cavitating jet, that is, the jet is impinging onto the sample surface. The chamber is filled with water in order to allow the occurrence of the bubbles implosion. The damaged area was measured, the pits number were counted and the samples were photographed every 60 seconds until 300 seconds of time test. Then, the test continued more 900 uninterrupted seconds, achieving a total test time of 1200 seconds.

Apparatus test parameters can be adjusted to obtain optimal efficiency. In this experiment, pressure and velocity of the system were adjusted to reach a cavitation index of 0.14, which is considered to cause damage in hydraulic structures [11]. The three different geometry nozzles were used to evaluate the efficiency of cavitating jets.

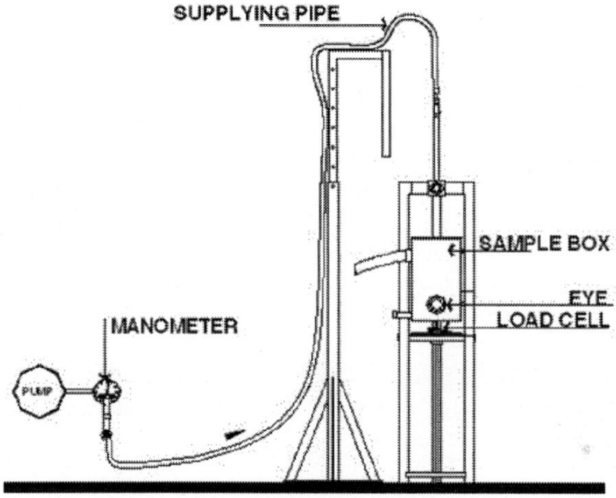

Figure 5. Schematic of experimental setup for erosion tests.

Concrete is a heterogeneous material, which poses difficulties to test and evaluate optimal conditions based on comparison test. Thus, aluminum samples were used to establish the nozzle efficiency because the metal is homogenous and allows a comparison among tests in order to evaluate the differing optimal erosion conditions, and damages. Cavitation erosion rates can be analyzed in terms of number of pits by time in order to obtain quantitative and qualitative information on the erosion intensity variation. A binocular magnifier 160x was used to count pits in the surface of the aluminum samples through the tests

First, the apparatus tests were run with the different nozzles presented in Figure 4. The typical value of α=1/4 used by [21] was adopted and maintained for comparison purposes. The sound velocity c in water at 20°C was assumed to be 1482 m/s and water density as 998.2 kg/m^3. Given the value of the jet velocity vj, calculations were done for all the data and the efficiencies η were calculated. It was presented by [7] the efficiency results for the nozzles (Table1).

Figure 6. Photo of experimental setup in the Laboratory.

Nozzle	Sharp-edge (ε)	Chamfered- edge (ε)
132°Conical	0.43	0.17
Circular	0.33	0.12
20°Conical	0.16	0.12

Table 1. Total erosion efficiency (ε) [7]

According to Table 1, better results were observed for sharp-edged nozzles than for chamfered-edge ones. The best total efficiency was obtained for the 132° conical nozzle (ε=0.43). Also, the time intervals required to accomplish the same erosion rate were observed to be 300 seconds for 132° conical sharp-edged nozzle, and 1200 seconds for the chamfered-edge nozzle. Therefore, not only the entrance pressure, but also nozzle geometry and finish are important for optimizing the apparatus efficiency. The authors pointed out that obtuse-angle nozzles with sharp-edge are preferable to perform cavitation erosion tests.

Using cavitating jet apparatus (σ=0.14) and the 132° conical sharp-edged nozzle, three special concrete samples were tested to obtain the cavitation erosion rate. The procedure of making samples begins by separating, cleaning and stocking the aggregates in the laboratory the day before. The next day, samples were molded in agreement with the Brazilian Standard NBR 5738. Table 2 presents samples compositions and compressive resistance obtained.

Sample	Cement and Aggregates (Relation)	Water and Cement (Relation)	Aggregate	Addition	Compressive Resistance (MPa)
A	1:4	0.3	Limestone	-	55.00
B	1:4	0.3	Granite	-	63.00
C	1:4	0.3	Granite	Silica *	83.00

* 8% the mass of cement

Table 2. Characteristics of the concrete samples.

Figure 7 shows the erosion comparison between concrete samples in terms of volume erosion in time.

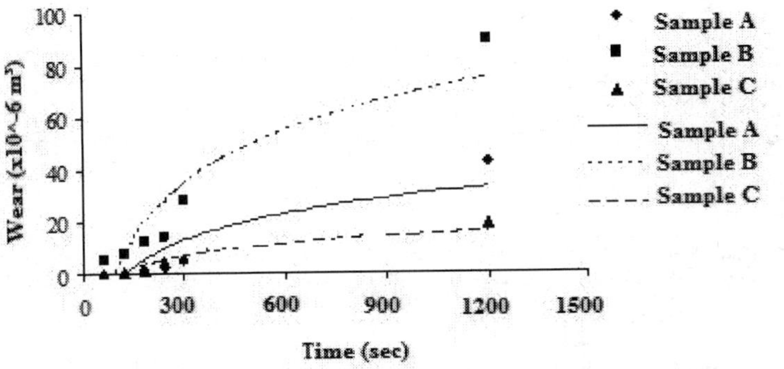

Figure 7. Concrete erosion over time.

The best results were obtained for sample C which contains hard aggregate, superior axial compressive resistance and the addition of silica. However sample A (fck 55.00 MPa) presented better results than sample B (fck 63.00 MPa), despite of the greater compressive resistance, showing that an adequate concrete (resistant to cavitation erosion) is a combination of several factors such as aggregate type, size and shape, additions to cement and water cement relation. Figure 8 shows the evolution of erosion in the B concrete sample. Additional tests were performed with unsubmerged and submerged samples, showing that the formation of a high-velocity cavitating jet provoked high erosion rates in the sample as expected, according [7]. In [24] the authors compared the erosion generated by the cavitating jet imping directly in to and parallel to a concrete sample. In this last case, the sample is not subjected to the effects of the impact force of the jet. Comparing the results, a higher erosion rate was observed when the samples are positioned directly over to the cavitating jet (Figure 9).

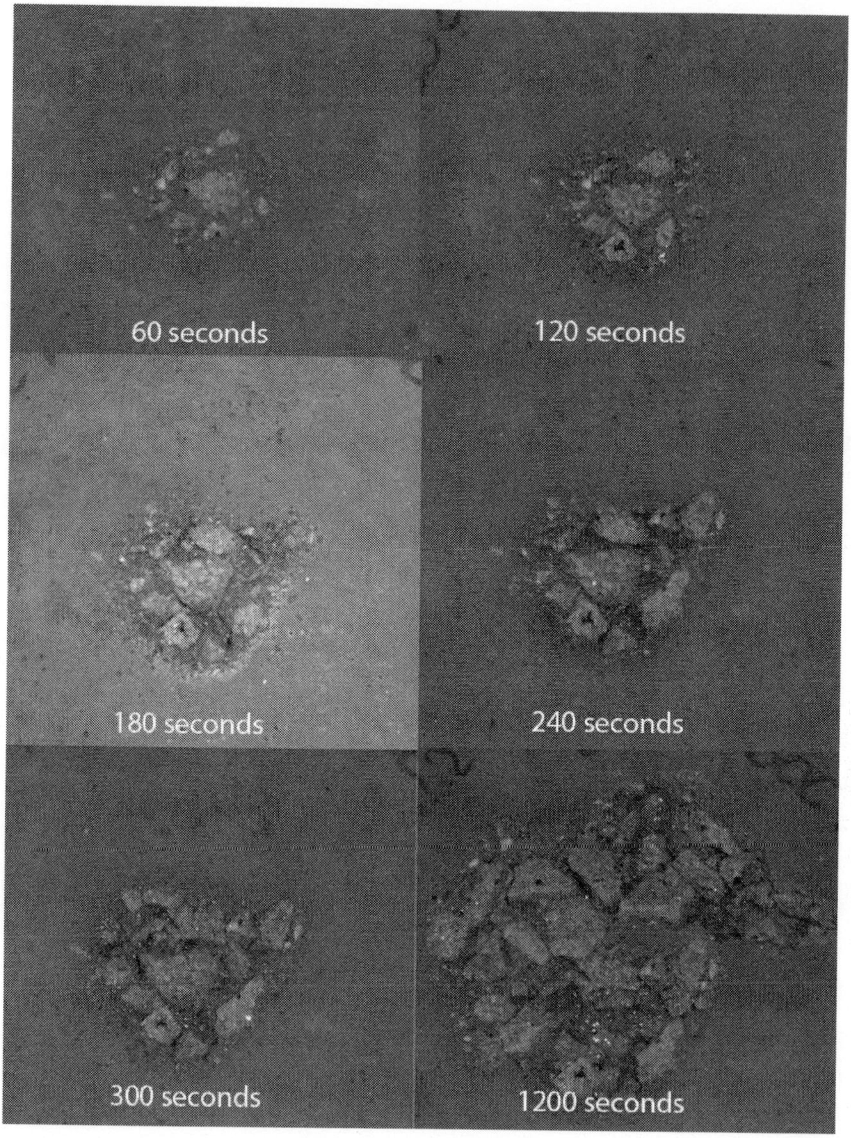

Figure 8. Erosion evolution of concrete sample B over time.

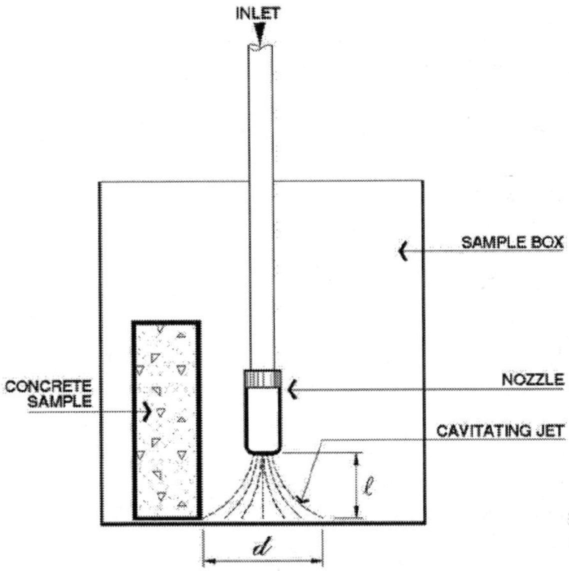

Figure 9. Concrete sample placed parallel to the cavitating jet [24].

Fast, efficient and economic testing is needed. The use of the cavitating jet allowed significant reduction in testing times, [10] specially when compared to the Venturi device. In [12] the author took 30 hours to compare the cavitating wear among concrete samples. As concrete is a heterogeneous material, samples are necessary to be larger than a few millimeters, as used by [25] to test metal samples. [10] used samples 20 cm in diameter and 5 cm in height to test concrete samples. The cavitating jet apparatus used in this experiment is compact, low cost and has short test times. It can also be used to measure erosion for testing concrete use in hydraulic structures. The cavitating jet apparatus creates a force larger than the one generated by a simple jet of high pressure, and thus can simulate the combined effect of high-speed flows and cavitation normally experienced in hydraulic structures

3. Use of the cavitating jet to inactivate bacteria and to decompose persistent compounds in water

3.1. Introduction

Water quality has deteriorated over the years due to industrial and agricultural pollution as well as an increase in the domestic sewage generated by a rapidly growing population. Pathogenic microorganisms in water pose a serious health and security threat to drinking water supply systems. The quality of water for human consumption can be improved by controlling pollution and by increasing the efficiency of inactivation techniques, which involves the destruction of pathogens present in water at a reasonable cost.

The use of chemicals, such as chlorine gas, sodium hypochlorite, calcium hypochlorite, hydrated ammonia, ammonium hydroxide, ammonium sulphate, and ozone, is common during the water disinfection process in conventional water treatment systems [26]. Many problems arise when using chemical methods for disinfection, including the high maintenance demands of the associated facilities (corrosion, incrustation), the formation of toxic byproducts (chlorine addition may generate byproducts such as trihalomethanes) and environmental concerns (chemical effluents released into rivers compromise aquatic life) [27]. Therefore, the development of alternative techniques for inactivating pathogens in water is desirable. Inactivation based on the cavitation phenomenon appears to be a promising alternative or supplement to existing techniques [28].

Research using the phenomenon of cavitation to inactivate microorganisms has been performed in the United States, Russia, India, China, Japan, UK, South Africa, France, Mexico, among others, using different methods. The hydrodynamic cavitation method is the most studied and disseminated. This is because the best results are achieved. However, there is currently no standardized method to carry out inactivation with this technique with respect to the pressures and speeds testing.

The performance shown by the studies on the wear of concrete indicated that the cavitating jet apparatus could be adapted to test the inactivation of microorganisms and the decomposition of persistent compounds in water. However, tests are necessary to establish appropriate pressures and velocities for each study. The inactivation of microorganisms was performed with *Escherichia coli* that are microorganisms commonly found in the intestines of humans and other warm-blooded animals and indicate the presence of water contamination. The decomposition of persistent compounds was performed with methylene blue.

3.2. Experimental tests for inactivation and chemical decomposition

As illustrated in Figure 2 the cavitating jet apparatus with 132° conical nozzle was adapted to conduct the experiments to inactivate *Escherichia coli* and to decompound the persistent methylene blue compound. The chamber used was 700.00×10^{-3} m high with a diameter of 300.00×10^{-3} m. Part of the container was filled with 40.00×10^{-3} m^3 of contaminated water to be studied in a closed circuit. A low-cost refrigeration system was adapted to the apparatus. A 12.70×10^{-3} m diameter and $19{,}500.00 \times 10^{-3}$ m long copper coil was placed inside of the inactivation chamber. This coil was connected to a reservoir with a capacity of 750.00×10^{-3} m^3 that was filled with clean water. The temperature of the tests was held constant at 33 °C, thus assuring that the inactivation was exclusively due to the cavitation phenomenon instead of an increase in the water temperature.

All of the microbiological, physical, and chemical procedures of this study follow the procedures described in the Standard Methods for Examination of Water and Wastewater Manual [19,20]. Non-pathogenic *Escherichia coli* (ATCC 25922) were used for the microbiological tests.

Viable cell counting was performed using the Colilert Method®. The presented results of the microbiological tests are the average of the results obtained in the dilutions.

The high-pressure displacement pump recirculated the water to be treated and was kept at a pressure of 4.00, 8.00, 10.00 and 12.00 MPa. First, the inactivation device was turned on without pressure to circulate the residue being treated. Samples were then collected at the initial time (T0) to obtain a control sample. Next, a vial of the sample was removed to measure the number of viable *Escherichia coli* cells. In addition to the control time point (T0), samples were collected every 900 seconds. All of the tests were repeated in triplicate, however, the results represent an average of the individual results.

Figure 10 shows that the higher the pressure, the higher the inactivation. At time test 1800 seconds, the following inactivation rates are observed: less than 37.50% for pressure test 4.00 MPa, 98.30% for test pressure 8.00 MPa, 99.96% for pressure test 10.00 MPa, and 100% for pressure 12.00 MPa,. Although the inactivation percentages are close to pressures 8.00, 10.00 and 12.00 MPa, the highest inactivation rate was achieved for test pressure 12.00 MPa.

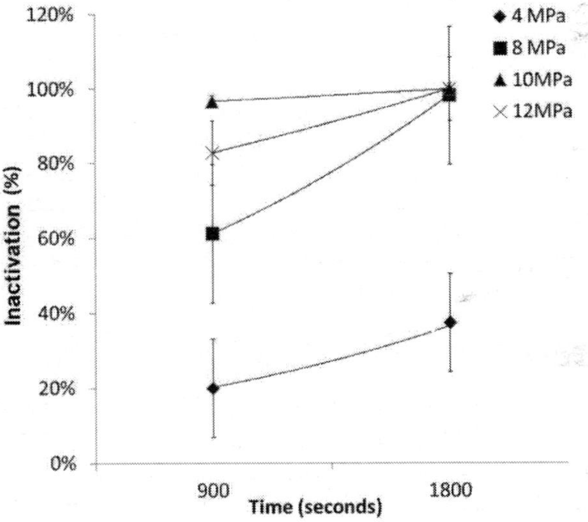

Figure 10. Inactivation rate and pressure.

However, the optimum inactivation rate will be achieved by relating the inactivation rate to the energy consumption. The Energy Efficiency (*EE*) of the apparatus is calculated with Equation 8 and the experiment results are shown in Table 3.

$$EE = \frac{Ci - Cf \cdot \forall}{P \cdot T} \tag{8}$$

being:

Cf: Final Concentration of *Escherichia coli* (CFU/mL);

Ci: Initial Concentration of *Escherichia coli* (CFU/mL);

∀: Volume (mL);

P: Power (W);

T: Time (seconds).

Parameters		Test pressures		
4.00 MPa	8.00 MPa	10.00 MPa	12.00 MPa	
t (s)	1800.00	1800.00	1800.00	1800.00
V (mL)	40000.00	40000.00	40000.00	40000.00
i (A)	25.00	29.00	31.00	32.80
U (V)	220.00	220.00	220.00	220.00
P (W)	5500.00	6380.00	6820.00	7216.00
Ci (CFU/mL)	918425.00	141360.00	800013.00	256547.00
Cf (CFU /mL)	574000.00	2410.00	337.00	9.00
Ci-Cf (CFU /mL)	344425.00	138950.00	799676.00	256538.00
(Ci-Cf)/P(CFU /mL)/(W)	62.62	21.78	117.25	35.55
€€ (CFU /J)	1391.62	483.98	2605.66	790.03

Table 3. Energy Efficiency to inactivate *Escherichia Coli* at different pressure.

In order to compare the tests, a time test of 1800 seconds was standardized. Table 3 indicates that for the conical 132° nozzle, the most efficient operation pressure is 10.00 MPa, followed by pressure 4.00, 12.00 and 8.00 MPa.

A complementary study was conducted with natural waters retrieved from a lake at the University's campus [28]. Given past studies by [29,30], high efficiency for the first 900 seconds of the test with an inactivation rate of 90% was obtained.

The substance methylene blue was used for testing degradation of persistent compounds. The results for each time and pressure test were analyzed using the spectrophotometer (Hach DR / 4000U Spectrophotometer) scan from 800 nm to 200 nm. To analyze the degradation of persistent compounds it is necessary to observe the decrease in absorbance at a given wavelength band over time. The absorbance is proportional to the concentration of the substance based on the law of Lambert-Bee. Changes in the molecule of methylene blue are then analyzed for two wavelength bands: the ranges from 200nm to 400nm, ultraviolet (UV), and the range from 400nm to 800nm, visible to the human eye.

The result of the degradation is shown in Figures 11 and 12 at a pressure of 12.00 MPa and time 1800 seconds. Figure 11 shows no significant reduction in absorbance. Figure 12 indicates a reduction of 10.5% in the final absorbance at the peak wavelength range of 664nm.

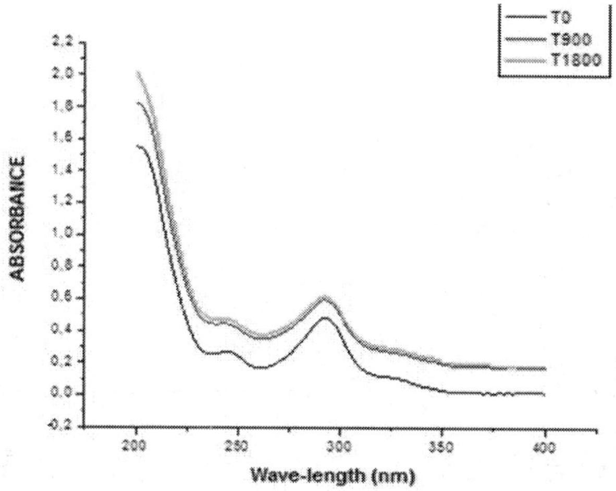

Figure 11. Decomposition of methylene blue, pressure 12 MPa, UV wavelength.

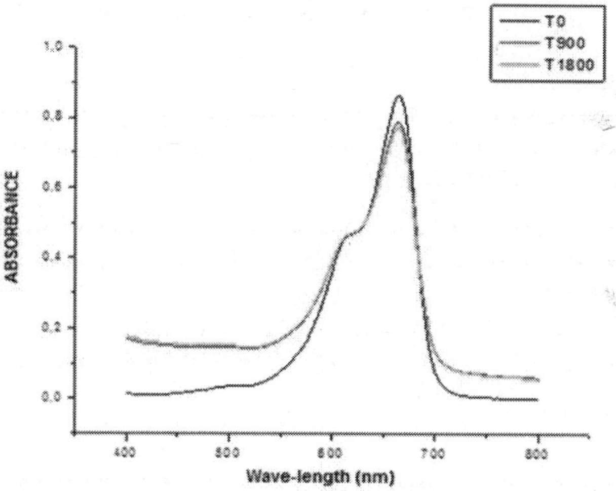

Figure 12. Decomposition of methylene blue, pressure 12 MPa, visible wavelength.

Comparing the results of Figures 11 and 12, the greatest difference in absorbance occurs in the visible band. Pressures lower than 12.00 MPa did not show any significant decrease in the absorbance. At the end of the tests, the residue was placed in translucent plastic bottles and exposed to sunlight. It was observed that the methylene blue dye was then degraded only by the peroxidation generated by the cavitation treatment.

The synergy between the cavitation process and the oxidative processes is a current and relevant research [31-33]. Recent work by [34-36] has been focusing on the degradation of persistent pollutants, such as pharmaceuticals, dyes, industrial effluents, pesticides, in addition to research with microorganisms.

4. Final remarks

Cavitation is known for its undesired effects produced in hydraulic systems, namely, the noise, vibrations, pressure fluctuations, erosion and efficiency loss. In hydraulic structures, cavitation can be destructive. Many times, the materials to do repairs in the structure are used without the necessary laboratory tests. These tests can be onerous, showing the importance of previous studies to check the applicability, either for the project phase or for repairs.

Despite the undesirable effects of cavitation, this phenomenon showed to be an interesting alternative to inactivate bacteria in water. The use of chemicals, such as chlorine, is commonly used for water disinfection in conventional water treatment systems. Cavitation can be used as a "clean" treatment, as it can reduce the quantity of chemicals added to water in the conventional water treatments.

The cavitating jet equipment simulates cavitation and its efficiency can be improved by using different nozzle configuration that forms the cavitating jet. This work showed that the most efficient nozzle geometry was obtained using a sharp-edge 132° conical nozzle for different applications, erosion tests, bacteria inactivation and decomposition of compounds.

The cavitating jet apparatus proved to be efficient, low cost, low power consumption when used to test material erosion and to inactivate *Escherichia coli*. Furthermore, it can be adapted to operate at larger scale treatments.

Acknowledgements

The authors would like to acknowledge FAPESP (São Paulo Research Foundation) for the Scientific Research Scholarship (Process: 2000/03611-0, 2000/03732-2, 2002/10348-0, 2009/53553-1 and 2011/16347-4) and Research Award Scheme (Process: 2009/54278-4).

http://dx.doi.org/10.5772/59136

Author details

José Gilberto Dalfré Filho*, Maiara Pereira Assis and Ana Inés Borri Genovez

*Address all correspondence to: dalfre@fec.unicamp.br

College of Civil Engineering, Architecture, and Urbanism, State University of Campinas, UNICAMP, Campinas, Brazil

References

[1] Billus, I., Predin, A., Skerget, L. (2007). The extended homogeneous cavitation transport model. J. Hydraulic Res. 45(1), 81-87.

[2] Wilhems, S.C., Gulliver, J.S. (2005). Gas transfer, cavitation, and bulking in self-aerated spillway flow. J. Hydraulic Res. 43(5), 532-539.

[3] Knapp, R.T.; Daily, J.W. and Hammitt, F.G. (1970) Cavitation. 1 ed. MacGraw-Hill. 578p.

[4] Dalfré Filho, J.G., Genovez, A.I.B. (2009). Alternative apparatus to evaluate cavitation damage. J. Mat. in Civ. Engrg. 21(12), 797-800.

[5] Vischer, D.L.; Hager, W.H. Dam hydraulics. New York: John Wiley & Sons, 1999. 305p.

[6] Tullis, J.P. Hydraulics of pipelines-pumps, valves, cavitation, transients. Ed. John Wiley and Sons: New York; 1989. 265p.

[7] Dalfré Filho, J.G., Genovez, A.I.B. (2008). Efficiency evaluation of cavitating jet apparatus. Proc. 2nd Int. Junior Researcher and Engineer Workshop on Hydraulic Structures, IAHR, Pisa, Italy, 107-112.

[8] Gomes, J., Marques, M., Matos, J. (2007). Predicting cavitation inception on steeply sloping stepped spillways. International Association of Hydraulic Research, IAHR, 32nd, 2, p.21.

[9] Dalfré Filho, J.G. (2002). Desenvolvimento de equipamento para avaliação de erosão por cavitação [Development jet device to evaluate erosion by cavitation]. Master Thesis in Civil Engineering, Department of Civil Engineering, Architecture and Urbanism, University of Campinas, São Paulo, Brazil.

[10] Dalfré Filho, J.G. (2005). Equipamento tipo jato cavitante para avaliação de desgaste por cavitação. [Cavitating jet device to evaluate erosion by cavitation]. Doctoral Thesis in Civil Engineering, Department of Civil Engineering, Architecture and Urbanism, University of Campinas, São Paulo, Brazil.

[11] Semenkov, V.M. 2007. Discussion on the work "Evaluating damage of concrete hydraulic structures caused by cavitation".

[12] Gikas, I. Cavitação-Efeitos sobre superfícies de resina de epóxi e concretos comuns e especiais. Boletim técnico DAEE, São Paulo,4(1), p.89-121, jan.-abril 1981.

[13] Billard, J.Y. and Fruman, D.H. (1992) "Etude expérimentale de i'influence d'un champ de pression fluctuant sur l'apparition de la cavitation dans un Venturi". La Houille Blanche, Paris, année 47, n.316, juillet/aout.

[14] Pham, T. M.; Michel, J.M.; Lecoffre,Y. (1997) "Qualification et optimization d'un compteur dynamique à ogive centrale". La Houille Blanche, Paris, année 52, n.342, p. 74-80, avril/mai.

[15] Momber, A.W. (2000) Short time cavitation erosion of concrete. WEAR, n.241, p. 47-52.

[16] Conn et al. Some unusual applications for cavitating water jets. In: International Symposium on Jet Cutting Technology, 7, 1984, Ottawa. Jet Cutting Technology. Cranfield: BHRA The Fluid Engineering Centre, june,1984. p.1-12.

[17] Cheng, C.L.; Webster, C.T. e Wong, J.Y. Cavitation resistant coatings for concrete hydraulic structures. ACI Materials Journal, Detroit, v.87, n.6, p.594-601, nov-dec1990.

[18] Momber, A.W. Deformation and fracture of rocks due to high-speed liquid impingement. International Journal of Fracture 130: 683–704, 2004.

[19] Momber, A.W. Aggregate liberation from concrete by flow cavitation. Int. J. Miner. Process., 74 (2002) 177-187.

[20] Hu, X.G.; Momber A.W.; Yin, Y. Erosive wear of hydraulic concrete with low steel fiber content. Journal of Hydraulic Enmgineering. ASCE.. p 1331-1340. Dec. 2006.

[21] Houlston, R., Vickers, G.W. (1978). Surface cleaning using water jet cavitation and droplet erosion. In: 4th Int. Sym. on Jet Cutting Technology. Cranfield: BHRA, The Fluid Engineering Centre, p.H1.1-H1.18.

[22] Kalumuck, K.M., Chahine, G.L., Hsiao, C.-T., Choi, J.-K. (2003). Remediation and disinfection of water using jet generated cavitation. Proc. 5th Int. Sym. Cavitation, Osaka, 1-8.

[23] Vickers, G.W., Houlston, R. Modelling the erosion efficiency of cleaning jets. Applied Scientific Research (40); 1983. 377-391.

[24] Dalfré Filho, J.G.; Genovez, A.I.B.; Luvizotto Junior (2006). Equipamento para Simulação da Erosão por Cavitação. Revista Brasileira de Recursos Hídricos, v.11, n.1, p. 51-62.

[25] Lichtarowicz, A. Cavitating jet apparatus for cavitating erosion testing. Erosion: prevention and useful applications. American Society for Testing and Materials, Philadelphia, STP 664, p. 530-549,1979.

[26] U.S. Environmental Protection Agency (1986). Technologies for upgrading existing or designing new drinking water treatment facilities. U.S. Environmental Protection Agency, Cincinnati.

[27] Koivusalo, M., Pukkala, E., Vartiainen, T., Jaakkola, J.J., Hakulinen, T. (1997). Drinking water chlorination and cancer—a historical cohort study in Finland. Cancer Causes Control. 8(2), 192-200.

[28] Assis, M.P.; Dalfré Filho, J.G.; Genovez, A.I.B. (2013). Bacteria Inactivation by means of Cavitation Blasting Equipment in Freshwater Systems. In: IAHR World 35th Congress. Anais...Chengdu, China.

[29] Assis, M.P.; Genovez, A.I.B.; Dalfré Filho, J.G. "Equipamento tipo jato cavitante para desinfecção de águas residuais" In: Simpósio Brasileiro de Recursos Hídricos, 2009.

[30] Assis, M.P.; Dalfré Filho, J.G., Genovez, A.I.B. "Equipamento tipo jato cavitante para degradação em compostos persistentes". XXIV Congreso Latinoamericano de Hidráulica, Punta del Este, Uruguay, nov.2010.

[31] Jyoti, K.K., Pandit, A.B. (2001). Water disinfection by acoustic and hydrodynamic cavitation. Biochem. Eng. J. 7(3), 201-212.

[32] Wang, X., Jia, J., Wang, Y. (2011). Degradation of C.I. Reactive Red 2 through photocatalysis coupled with water jet cavitation. J. Hazard. Mater. 185(1), 315-321.

[33] Braeutigam, P. et al. Degradation of carbamazepine in environmentally relevant concentrations in water by Hydrodynamic-Acoustic-Cavitation (HAC). Water Research. v.46, pp. 2469-2477, 2012.

[34] Wang, X.; Zhang, Y. Degradation of alachlor in aqueus solution by using hydrodinamic cavitation. Journal of Hazardous Materials, n. 161, pp. 202-207, 2009.

[35] Saharan, V. K.; Badve, M. P.; Pandit, A. B. Degradation of reactive Red 120 dye using hydrodynamic cavitation. Chemical Engineering Journal, n. 178, pp. 100-107, 2011.

[36] Wang, X.; Zhang, Y. Degradation of alachlor in aqueus solution by using hydrodinamic cavitation. Journal of Hazardous Materials, n. 161, pp. 202-207, 2009.

Chapter 4

Details of Hydraulic Jumps for Design Criteria of Hydraulic Structures

Harry Edmar Schulz, Juliana Dorn Nóbrega,
André Luiz Andrade Simões, Henry Schulz and
Rodrigo de Melo Porto

Additional information is available at the end of the chapter

http://dx.doi.org/10.5772/58963

1. Introduction

The sudden transition from supercritical to subcritical flow, known as hydraulic jump, is a phenomenon that, although being studied along decades, still presents aspects that need better quantification. Geometrical characteristics, such as the length of the roller (or the jump itself), still have no definitive formulation for designers of hydraulic structures. Even predictions of the sequent depths, usually made considering no shear forces, may present deviations from the observed values.

In the present chapter the geometrical characteristics of hydraulic jumps are obtained following different deductive schemes. Firstly, two adequate control volumes and the principles of conservation of mass, momentum and energy were used to obtain the length of the roller and the sequent depths. The conditions of presence or absence of bed shear forces are discussed. Secondly, two ways are used to propose the form of surface profiles: i) a "depth deficit" criterion and ii) the mass conservation principle using an "air capture" formulation. The presence or absence of inflexion points is discussed considering both formulations. Finally, the height attained by surface fluctuations (water waves and drops), useful for the design of the lateral walls that confine the jumps, is considered using empirical information and an approximation based on results of the Random Square Waves method (RSW).

Experimental data from the literature were used for comparisons with the proposed theoretical equations, allowing the adjustment of coefficients defined in these equations.

1.1. General aspects

A supercritical flow in a long horizontal channel (or having a small slope) has an inexpressive component of the weight as impulsive force. As a result, the power dissipation induces the depth of the liquid to increase along the flow, tending to its critical value. But a sudden change, a hydraulic jump, forms before the flow reaches the critical condition (which corresponds to the condition of minimum specific energy). In other words, transitions to the subcritical condition do not occur gradually in such channels. Hydraulic jumps have been studied over the years, providing a good theoretical understanding of the phenomenon, as well as experimental data that can be used to check new conceptual proposals.

The characteristics of hydraulic jumps in free flows are of interest for different applications. For example, they are used in structures designed for flow rate measurements, or for energy dissipation, like the stilling basins downstream from spillways. Although simple to produce and to control in the engineering praxis, hydraulic jumps still present aspects which are not completely understood. Interestingly, the geometrical characteristics of the jumps are among the features whose quantification still not have consensus among researchers and professionals. Perhaps the length of the jump (or the roller) is the dimension that presents the highest level of "imprecision" when quantifying it. Such difficulties may be a consequence of the simplifying hypotheses made to obtain the usual relationships. In the most known equation for the sequent depths, the shear force is neglected, and a very simple functional dependence is obtained between the depths and the Froude number (*Fr*). However, by neglecting the shear force, difficulties arise to calculate the length of the jump, because it does not appear in the formulation.

Many authors observed this difficulty since the beginning of the studies on hydraulic jumps. As a consequence, also a large number of solutions were presented for the sequent depths and for the length of the jump (or roller). Good reviews were presented, for example, by [1] and [2].

The hydraulic jump may be viewed as a "shock" between the supercritical and the subcritical flows. As result of this "shock", the water of the higher depth "falls down" on the water of the lower depth. Because of the main movement of the flow, and the resulting shear forces, a roller is formed in this "falling region", characterized as a two phase turbulent zone with rotating motion. This description is sketched approximately in Figure 1a. Figure 1b shows an example of this highly turbulent region, responsible for most of the energy dissipation in a hydraulic jump. Despite the fact of being turbulent (thus oscillating in time) and 3D, the hydraulic jump may be quantified as a permanent 1D flow when considering mean quantities.

Table 1 presents some studies related to the theme around the world, while Table 2 lists some academic studies in Portuguese language, developed in universities of Brazil, as a consequence of the increasing interest on hydroelectric energy in the country, and the corresponding projects of spillways with dissipation basins. The present chapter uses the background furnished by the previous studies.

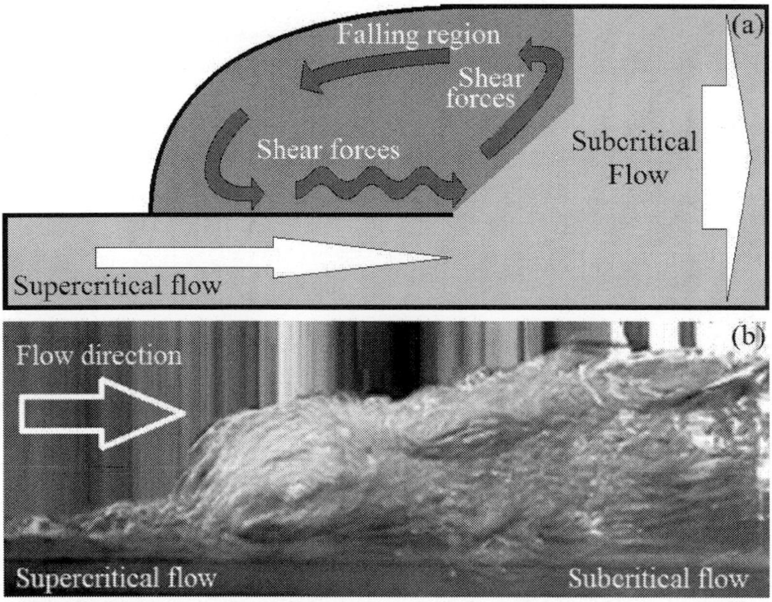

Figure 1. a) The "shock" between supercritical and subcritical flows, and the "falling region" forming the roller; b) An example of the "shock" sketched in Figure 1a.

Author (quoted by [1] and [2])	Contribution
Bidone (1818)	Experimental observations.
Belanger (1828)	Second Newtons' law applied to the jump.
Darcy & Bazin (1856/58, published 1865)	Experimental study for the evaluation of the subcritical height.
Bresse (1860)	Second Newtons' law applied to the jump.
Unwin (1880)	Second Newtons' law applied to the jump.
Feriday (1894)	Experimental study.
Merriman (1903)	Second Newtons' law applied to the jump.
Gibson (1913)	Verifying the Newtons' law.
Kennison (1916)	Second Newtons' law applied to the jump.
Riegel & Beebe (1917)	Description of the phenomenon. Verifying the Newtons' law. Discussion of the length of the jump.
Safranez (1927, 1929)	Verifying Newtons' law. Discussion of the length of the jump.
Knapp (1932)	Length of the jump.
Pietrkowski (1932)	Length of the jump.
Trahern (1932)	Localization of the jump.
Inglis & Joglekar (1933)	Second Newtons' law applied to the jump.

Author (quoted by [1] and [2])	**Contribution**
Einwachter (1933)	Length of the jump.
Ludin & Barnes (1934)	Length of the jump.
Woycieki (1934)	Length of the jump.
Smetana (1934)	Length of the jump.
Douma (1934)	Length of the jump.
Aravin (1935)	Length of the jump.
Kinney (1935)	Experiments and Length of the jump.
Page (1935)	Length of the jump.
Chertoussov (1935)	Length of the jump.
Bakhmeteff & Matzke (1936)	Extensive project for the generalization of equations.
Ivanchenko (1936)	Length of the jump.
Willes (1937)	Length of the jump and air entrainment.
Goodrum & Dubrow (1941)	Influence of surface tension and viscosity.
Posey (1941)	Length of the jump.
Wu (1949)	Length of the jump.
Bureau of Reclamation (1954)	Experimental study.
Schröder (1963)	Length of the jump.
Rajaratnam (1965)	Length of the jump.
Malik (1972)	Length of the jump.
Sarma & Newnham (1973)	Length of the jump.
Hager *et al.* (1990) [2]	Length of the roller.

Table 1. Studies related to hydraulic jumps. Table considering information contained in the studies of [1] and [2].

Having so many proposals, it may be asked why is there no consensus around one of the possible ways to calculate the geometrical characteristics. A probable answer could be that different solutions were obtained by detailing differently some of the aspects of the problem. But perhaps the most convincing answer is that the presented mechanistic schemes are still not fully convincing. So, it does not seem to be a question about the correctness of the used concepts, but on how they are used to obtain solutions. In other words, it seems that there are still no definitive criteria, generally accepted.

This chapter shows firstly the traditional quantification of the sequent depths of hydraulic jumps, which does not consider the bed shear force. In the sequence, using two adequate control volumes and the principles of conservation of mass, momentum and energy, an equation for the length of the roller is furnished. Further, the sequent depths are calculated considering the presence and the absence of bed shear forces, showing that both situations lead to different equations. Still further, the surface profiles and the height attained by the surface fluctuations (drops and waves) are discussed, and equations are furnished using i) a "depth deficit" criterion, ii) the mass conservation principle using an "air capture" formulation, and iii) an approximation that uses results of the Random Square Waves method (RSW). Literature data are used to quantify coefficients of the equations and to compare measured and calculated results.

Author	Advisor	Nature of study	University
Jayme Pinto Ortiz (1982) [3]	Milton Spencer Veras Jr.	MSc Dissertation	USP*
Jayme Pinto Ortiz (1989) [4]	Angelo Raffaele Cuomo	Dr Thesis	USP
Margarita M. Lopez Gil (1991) [5]	Kokei Uehara	MSc Dissertation	USP
Yosuke Yamashiki (1994) [6]	Podalyro Amaral de Souza	MSc Dissertation	USP
Jaime Frederici Gomes (2000) [7]	Marcelo Giulian Marques	MSc Dissertation	UFRGS**
Edgar F. Trierweiler Neto (2006) [8]	Marcelo Giulian Marques	MSc Dissertation	UFRGS
Rafael André Wiest (2008) [9]	Marcelo Giulian Marques	MSc Dissertation	UFRGS
Alexandre A. Mees Alves (2008) [10]	Marcelo Giulian Marques	MSc Dissertation	UFRGS
Daniela Müller de Quevedo (2008) [11]	Robin Thomas Clarke	Dr Thesis	UFRGS
Simone Maffini Cerezer (2008) [12]	Robin Thomas Clarke	Dr Thesis	UFRGS
Eder Daniel Teixeira (2008) [13]	Marcelo Giulian Marques	Dr Thesis	UFRGS
Jayme Pinto Ortiz (2011) [14]	-	Assoc. Prof. Thesis	USP
Maurício Dai Prá (2011) [15]	Marcelo Giulian Marques	Dr Thesis	UFRGS
Pedro E. de A. e Souza (2012) [16]	Marcelo Giulian Marques	MSc Dissertation	UFRGS
Juliana Dorn Nóbrega (2014) [17]	Harry Edmar Schulz	MSc Dissertation	USP

Table 2. Academic studies conducted in Brazil about hydraulic jumps. *USP: University of São Paulo; **UFRGS: Federal University of Rio Grande do Sul.

1.2. Traditional quantification of sequent depths

The Newton's second law for a closed system establishes that:

$$\sum \vec{F} = d\left(m\vec{V}\right) / dt\,|_S \tag{1}$$

where $\vec{F}$ is the force, m is the mass of the system, $\vec{V}$ is the speed, and t is the time. The system is indicated by the index S. Considering the Euler formulation, for open systems or control volumes, equation (1) is rewritten using the Reynolds transport theorem, furnishing:

$$\sum \vec{F} = \frac{d}{dt}\iiint_{VC} \vec{V}\rho dVol + \oiint_{SC} \vec{V}\left(\rho\vec{V}\cdot\vec{n}dA\right) \tag{2}$$

where ρ is the density, $\vec{n}$ is the unit vector normal to the control surface and pointing outside of the control volume, and A is the area of the control surface.

Hydraulic jumps in horizontal or nearly horizontal channels are usually quantified using a control volume as shown in Figure 2, with inlet section 1 and outlet section 2, no relevant shear forces at the bottom and the upper surface (F_{shear}), no effects of the weight (W_x), and hydrostatic pressure distributions at sections 1 and 2.

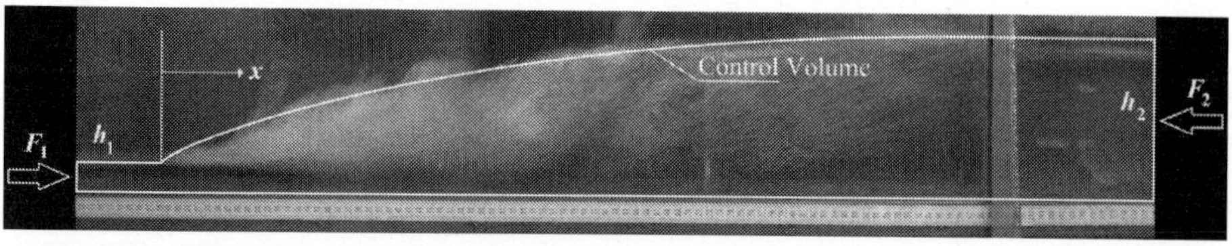

Figure 2. Hydraulic jump and adopted control volume. The photograph was taken to show the air bubbles ascending, and the outlet section located where no air bubbles are present.

For steady state conditions, equation (2) is written as:

$$\{F_1 - F_2 + W_x - F_{shear} \cong F_1 - F_2 =\} \quad \sum \vec{F} = \oiint_{SC} \vec{V}\left(\rho \vec{V} \cdot \vec{n} dA\right) \quad \{= \rho\left(V_2^2 A_2 - V_1^2 A_1\right)\} \tag{3}$$

In the sequence, from $Q=V_1A_1=V_2A_2$:

$$F_1 - F_2 = \rho Q^2 \left(\frac{1}{A_2} - \frac{1}{A_1}\right) \tag{4}$$

Q is the water flow rate, ρ is the water density, F_1 and F_2 are the forces due to the pressure at the inlet and outlet sections, respectively. The hydrostatic distribution of pressure leads to:

$$F_1 = \rho g h_1 A_1 / 2 \qquad \text{and} \qquad F_2 = \rho g h_2 A_2 / 2 \tag{5}$$

g is the acceleration of the gravity, h_1 and h_2 are the water depths, and A_1 and A_2 are the areas of the cross sections 1 and 2, respectively. Substituting equations (5) into equation (4) produces:

$$Q^2 / gA_1 + A_1 h_1 / 2 = Q^2 / gA_2 + A_2 h_2 / 2 \tag{6}$$

For a rectangular channel $A_1=Bh_1$, $A_2=Bh_2$, where B is the width of the channel. Further, the definition $q=Q/B$ is of current use. Algebraic steps applied to equation (6) lead to:

$$h_1^2 - h_2^2 = \frac{2q^2}{g}\left(\frac{1}{h_2} - \frac{1}{h_1}\right) \Rightarrow (h_2 - h_1)(h_2 + h_1) = \frac{2q^2}{g}\left(\frac{1}{h_2} - \frac{1}{h_1}\right) \Rightarrow (h_2 + h_1) = \frac{2q^2}{gh_1h_2} \Rightarrow$$

$$\Rightarrow h_1^2 h_2 + h_2^2 h_1 = \frac{2q^2}{g} \Rightarrow \frac{h_2^2}{h_1^2} + \frac{h_2}{h_1} - \frac{2q^2}{gh_1^3} = 0 \quad \text{or} \quad \left(\frac{h_2}{h_1}\right)^2 + \frac{h_2}{h_1} - 2Fr_1^2 = 0$$

http://dx.doi.org/10.5772/58963

Solving the final obtained equation for h_2/h_1, the result is:

$$h^{\#} = h_2 / h_1 = \left(\sqrt{8Fr_1^2 + 1} - 1\right) / 2 \quad \text{where} \quad Fr_1 = V_1 / \sqrt{gh_1} \tag{7}$$

Figure 3 shows that equation (7) compares well with experimental results.

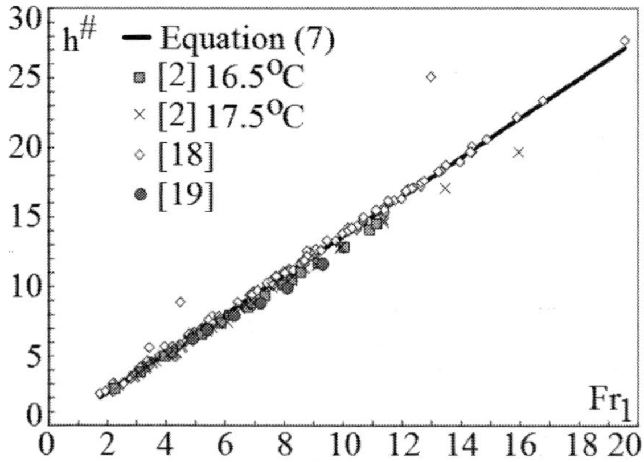

Figure 3. Sequent depths ratio as a function of the Froude number. Data of [2, 18, 19].

The experimental data of [2] and [19] show that predictions of equation (7) may generate relative errors in the range of 0.10% to 12.2%. Considering simpler equations, different proposals may be found in the literature. For example. for $Fr_1>2$ [20] suggests the simplified form $h^{\#}=\sqrt{2}Fr_1-1/2$, (where $h^{\#}=h_2/h_1$). But other linear equations are also suggested. Equation (8), for example, is valid for $2.26<Fr_1<15.96$, has a correlation coefficient of 0.99, maximum relative error of 5.4%, for the data of Figure 3, excluding [18].

$$h^{\#} = 1.29Fr_1 - 0.116 \tag{8}$$

Although alternative equations exist in the literature, equation (7) is one of the most known and accepted conclusions for hydraulic jumps. Interestingly, hydraulic jumps are used as dissipative singularities, but their most known design equation implies absence of dissipative shear forces. This contradiction is one of the reasons of the continuing discussion on the theme. In this chapter we present (see items 2.1.1, 2.1.2, 2.1.3) a way to conciliate the dissipative characteristic of the jump and the adequacy of the predictions given by equation (7).

1.3. Equations of the literature for the lengths of the roller and the hydraulic jump

The experimental determination of the length of the hydraulic jump is not a simple task. The intense turbulence and the occurrence of single-phase and two-phase flows adds difficulties to the measurement of flow depths, velocity fields, pressure distributions and the lengths of the roller and the hydraulic jump. A further difficulty is related to the definition of the end of the hydraulic jump. Accordingly to [1], the earliest formulation for the length of the hydraulic jump was proposed by, Riegel and Beebe, in 1917, while [2] suggest that the first systematic study of the length of the roller was made by Safranez, in 1927-1929. [21] defined the end of the hydraulic jump as the position where the free surface attains its maximum height, and the upper point of the expanding main flow (located between the roller and the bottom of the channel) coincides with the surface, beginning to decline towards the subsequent gradually varied flow. This definition led to lengths greater than the rollers.

[2] mention Schröder, who in 1963 used visualizations of the free surface to quantify jump lengths. However, such visual procedures depend on personal decisions about different aspects of the moving and undulating surface. [2] also cite Malik, who in 1972 employed a probe to measure forces and to locate the superficial region with zero mean force. The position of this region corresponds to the roller length, and the experimental error was about 8%.

The results of [18] are among the most used for calculating lengths of hydraulic jump stilling basins, downstream of spillways. The end of the jump was assumed as the position where the high velocity jet starts to peel off the bottom, or the section immediately downstream of the roller.

[19] measured pressure distributions along the bottom of a horizontal channel for hydraulic jumps with Fr_1 between 4.9 and 9.3. The pressure records were used to calculate coefficients of skewness and kurtosis of the measured data along the channel. The authors obtained values of coefficients of kurtosis around 3.0, which became practically constant for distances greater than $x/(h_2\text{-}h_1)$=8.5. The authors then defined the end of the hydraulic jump as L_j=8.5(h_2-h_1).

[22, 23] measured water depths along a hydraulic jump in a rectangular channel, for Fr_1=3.0, using an ultrasonic sensor to locate the water surface. The sensor was moved along the longitudinal axis of the channel. Vertical turbulent intensities and related Strouhal numbers were calculated for each measurement position. The authors suggested to estimate the length of the jump using the final decay of the vertical turbulent intensity at the free surface, obtaining L_j=9.5(h_2-h_1). Similarly, [24, 25] and [17] generated and analyzed data of ultrasonic sensors and high speed cameras to evaluate comparatively the results of the sensors and to better locate the surface.

Some equations for the length of the hydraulic jump and the roller are resumed in Table 3. Most of them can be written as L_j/h_2=f(Fr_1) when using equation (7). A qualitative comparison between some of these equations is shown in Figure 4, assuming the interval $2 \le Fr_1 \le 20$ as valid for all equations.

Riegel and Beebe (1917):	$L_j \approx 5(h_2 - h_1)$	(9)
Safranez (1927):	$L_j \approx 5.2h_2$	(10)
Ludin and Barnes (1934):	$L_j = (4.5 - V_1 / V_c)h_2$	(11)
Woycieki (1934):	$L_j = (h_2 - h_1)(8 - 0.05h^{\#})$	(12)
Smetana (1934):	$L_j \approx 6(h_2 - h_1)$	(13)
Douma (1934):	$L_j = 5.2h_2$	(14)
Aravin (1935):	$L_j \approx 5.4(h_2 - h_1)$	(15)
Kinney (1935):	$L_j = 6.02(h_2 - h_1)$	(16)
Page (1935):	$L_j = 5.6h_2$	(17)
Chertoussov(1935):	$L_j = 10.3h_1(Fr_1 - 1)^{0.81}$	(18)
Bakhmetef, Matzke(1936):	$L_j = 5(h_2 - h_1)$	(19)
Ivanchenko (1936):	$L_j = 10.6(h_2 - h_1)(Fr_1^2)^{-0.185}$	(20)
Posey (1941):	$L_j \approx 4.5 - 7(h_2 - h_1)$	(21)
Wu (1949):	$L_j = 10(h_2 - h_1)Fr_1^{-0.16}$	(22)
Hager *et al.*(1992) [20]:	$L_j / h_1 = 220tgh[(Fr_1 - 1)]$	(23)
Hager *et al.* (1990) [2]:	$L_r / h_1 = 12 + 8a_r tgh(Fr_1 / a_r)^{\$}$	(24)
Marques *et al.* (1997) [19]:	$L_j = 8.5(h_2 - h_1)$	(25)
Simões *et al.* (2012) [23]:	$L_j = 9.52(h_2 - h_1)$	(26)
Knapp (1932):	$L_j = (62.5h_1 / E_1 + 11.3)\left[(V_1 - V_2)^2 / (2g) - (E_1 - E_2)\right]$	(27)
Einwachter (1933):	$L_r = (15.2 - 0.24h^{\#})[(h^{\#} - 1) - V_1^2(h^{\#} - 1)/(h^{\#}2g)$	(28)
Simões (2008) [26]:	$L_j / h_2 = (Fr_1^2 - 81.85Fr_1 + 61.13)/(-0.62 - 10.71Fr_1)$	(29)

Table 3. Equations for the length of the roller and the hydraulic jump. The table uses citations of [1, 2, 18]. V_c=critical velocity; E=specific energy. $^{\$}\alpha_r$=20 if h_1/B<0.1; α_r=12,5 if 0.1≤h_1/B≤0.7. B=channel width. If Fr_1<6, following approximation may eventually be used: L_r/h_1=8Fr_1-12.

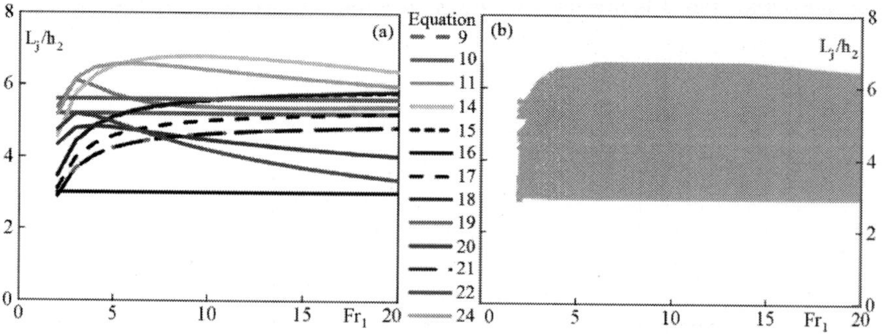

Figure 4. a) Comparison of equations of nondimensional jump lengths proposed by different authors cited in Table 3, as a function of Fr_1; b) The region of the graph covered by the equations.

Experimental data of [18] for L_j/h_2, conducted for Fr_1 between about 2 and 20, are shown in Figure 5. The middle curve corresponds to the adjusted equation (29) proposed by [26].

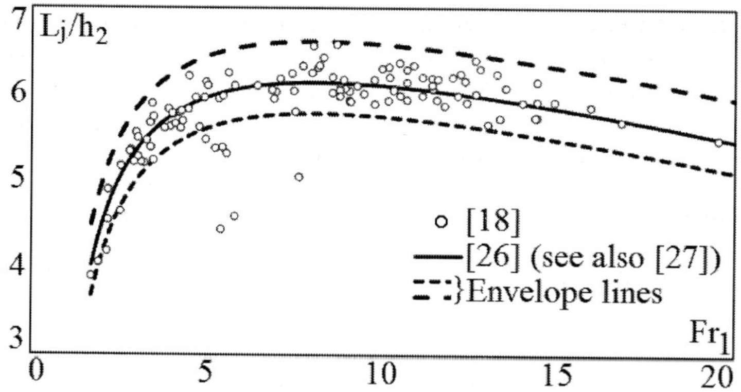

Figure 5. Data of L_j/h_2 obtained by [18], and equation (29) presented by[26] (see also [27]). The upper and lower curves are shown as envelopes.

2. Proposed formulation

2.1. The criterion of two control volumes

The focus of this item is to present the equations obtained for the length of the roller and for the sequent depths using only one deduction schema. A detailed explanation of the forces that maintain the equilibrium of the expansion is presented firstly, followed by their use to quantify the power dissipation. The integral analysis is used for the two control volumes shown in Figure 6.

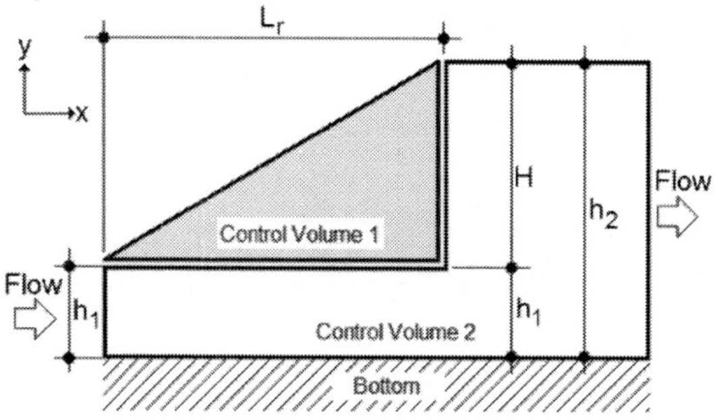

Figure 6. The two control volumes used in the present formulation, for the quantification of the length of the roller and the sequent depths of the hydraulic jump.

The two control volumes allow considering details, such as the vortex movement of the roller and the recirculating flow rate. The regime of the flow is permanent (stationary conditions), and both control volumes are at rest. In order to simplify the calculations, control volume 1 (CV1) contains the roller of the jump and ideally does not exchange mass with control volume 2 (CV2). But, as a result of the shear forces at the interface between both volumes, they exchange momentum and energy. The analyses of the flows in both volumes are made here separately.

2.1.1. Length of the roller using control volume 1 (CV1)

i. Global Analysis of CV1

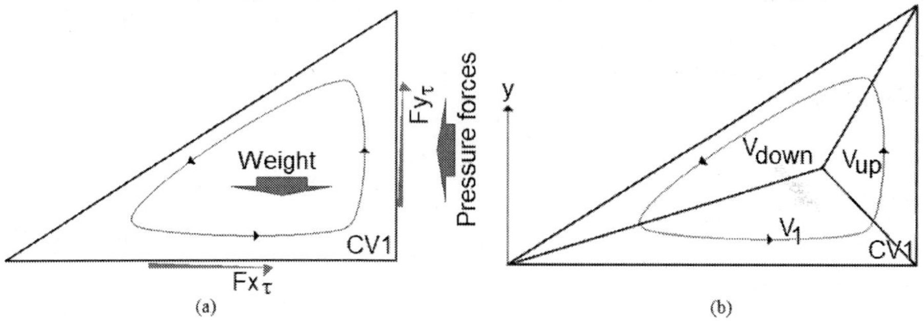

Figure 7. a) CV1 and the forces acting on it. $F_{x\tau}$ and $F_{y\tau}$ are shear forces acting in the x and y directions, respectively; b) The three regions of influence of the characteristic velocities (V_1, V_{up}, and V_{down}) defined for CV1.

Figure 7a shows CV1 already isolated from CV2. Because there is no mass exchange between CV1 and the main flow (CV2), the mass conservation equation (30) reduces to the elementary form (31):

$$\frac{d}{dt}\iiint_{CV1} \rho dVol + \oiint_{CS} \rho\vec{V}.d\vec{A} = 0 \tag{30}$$

$$\frac{dM}{dt} = 0 \tag{31}$$

CS is the control surface and *M* is the mass in CV1. For the momentum conservation, the general integral equation also reduces to the simplest form:

$$\vec{F} = \frac{d}{dt}\iiint_{CV1} V\rho dVol + \oiint_{CS} \vec{V}\rho\vec{V}.d\vec{A} = 0 \tag{32}$$

Because there are no mass fluxes across the surfaces, and the flow is stationary, the resultant of the forces must vanish. It implies that, for the coordinate directions x and y:

$$F_x = 0 \qquad \text{and} \qquad F_y = 0 \tag{33}$$

The mass in CV1 is constant and the volume does not change its form (that is, it will not slump and flow away). It implies that the shear forces of the flow must equate the pressure forces in the x direction, and the weight in the y direction. So, from Figure 7a, and equations (33) it follows that:

$$F_{x\tau} = \frac{\rho g H^2 B}{2} \tag{34}$$

$$F_{y\tau} = \frac{\rho g L_r B H}{2}\theta_1 \tag{35}$$

θ_1 is a constant that corrects the effects of using the inclined straight line instead of the real form of the water surface (understood as the water equivalent depth). H=h_2-h_1 and L_r is the length of the roller.

The shear forces presented in Figure 7a induce movement to the water in CV1, so that, in a first step, power is inserted into CV1. No mass exchanges occur across the control surface, and this energy is converted into thermal energy in CV1. But, further, this thermal energy is lost to the environment (main flow and atmosphere), so that the equilibrium or the stationary situation is maintained, and the energy inserted into the volume is released as heat to the environment.

Considering the energy equation, for stationary conditions and no mass exchange between the control volumes, it follows that:

$$\dot{Q}-\dot{W}=\frac{d}{dt}\iiint_{CV1} e\rho dVol+\oint\!\!\oint_{CS}(e+\frac{p}{\rho})\rho\vec{V}.d\vec{A}=0 \qquad \text{where} \qquad e=\frac{V^2}{2}+gy+u \tag{36}$$

$\dot{W}$ and $\dot{Q}$ are the work and the thermal energy exchanged with the surroundings per unit time, u is the specific internal energy and y is shown in Figure 7b. The surface integral vanishes because there is no mass flow across the control surface. From the first member of equation (36) it follows that:

$$\dot{W}=\dot{Q}=Power\ loss \tag{37}$$

Equation (37) means that, in this stationary case, all power furnished to CV1 is lost as heat. Equations (34, 35) and (37) allow writing:

$$\frac{\rho g H^2 BV_1}{2}+\frac{\rho g LBH}{2}\left(\frac{H}{\Delta t_y}\right)_{up}\theta_1=Power\ loss \tag{38}$$

$(H/\Delta t_y)$ is the mean vertical velocity (V_{up} in Figure 7b) that allows calculating the power furnished by the upwards force. Δt_y is the mean travel time of the water in the vertical direction (positive y axis, see Figure 7b). The mean upwards flow rate equals the mean downwards flow rate (because no mass is lost in CV1), so that, considering mean quantities for the involved areas, it follows that:

$$Area_{up}\left(\frac{H}{\Delta t_y}\right)_{up}=Area_{down}\left(\frac{H}{\Delta t_y}\right)_{down} \tag{39}$$

Or, because H is the same for both directions (up and down):

$$\Delta t_{yup}=\left(\frac{Area_{up}}{Area_{down}}\right)\Delta t_{ydown} \tag{40}$$

Equations (31) through (40) were obtained from integral conservation principles. In order to quantify the geometrical characteristics of the hydraulic jump, the variables *Power loss* and $\Delta t_{y_{up}}$ still need to be expressed as functions of the basic flow parameters. It is necessary, now, to consider the movement inside of the CV1, that is, to perform an intrinsic analysis in addition to the global analysis.

ii. Intrinsic Analysis of CV1

The total power loss in CV1 is the result of losses occurring in regions subjected to different characteristic velocities. Easily recognized are the velocity V_1 and the upwards velocity $H/\Delta t_{y_{up}}$. Furthermore, from equation (40) it follows that $\Delta t_{y_{up}}$ can be obtained from the downwards movement. From these considerations, three regions influenced by different "characteristic velocities" may be defined in CV1, as shown in Figure 7b. The *Power loss* of equation (38) is thus calculated as a sum of the losses in each region of Figure 7b, that is:

$$Power\,loss = \dot{W}\left(V_1\right)+\dot{W}\left(V_{up}\right)+\dot{W}\left(V_{down}\right) \tag{41}$$

The downwards region is considered firstly, following a particle of fluid moving at the "falling" boundary of the control surface, as shown in Figure 8a. (See also Figures 1a and 1b).

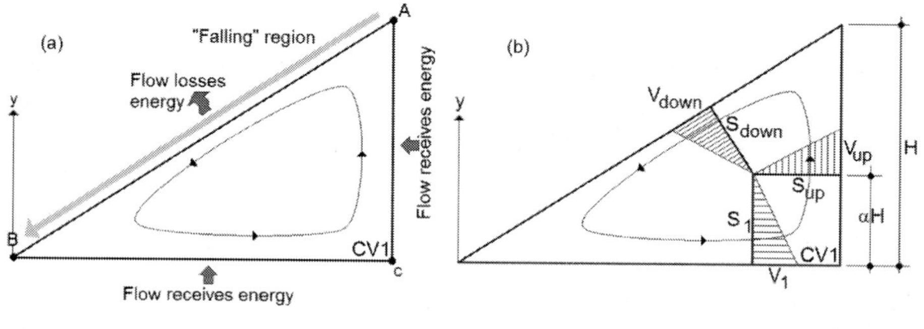

Figure 8. a) The "falling" of a particle in the "downwards region", between points A and B; b) Transversal sections S_1, S_{up}, and S_{down} in the roller.

Taking the movement along $\overline{AB}$ in Figure 8a, equation (42) applies:

$$y_A+\frac{p_A}{\rho g}+\frac{V_A^2}{2g}=y_B+\frac{p_B}{\rho g}+\frac{V_B^2}{2g}+h_{loss} \tag{42}$$

p is the pressure, which does not vary at the surface. V_A equals zero, so that the characteristic velocity is a function of V_B. Considering V_B=V_{down} and y_B=0, it follows:

$$H = \frac{V_{down}^2}{2g} + h_{loss} \tag{43}$$

For local dissipations h_{loss} is represented as proportional to the kinetic energy, considering a representative velocity. For the region of V_{down} (see Figures 7b and 8) it follows that:

$$h_{loss} = K\frac{V_{down}^2}{2g}, \qquad V_{down} = \sqrt{\frac{2gH}{1+K}} = \sqrt{2g'H}, \qquad g' = \frac{g}{1+K} \tag{44}$$

The "falling time" along $\overline{AB}$ is thus given by:

$$\Delta t_{y\,down} = \sqrt{\frac{2H}{g'}} \tag{45}$$

Equations (40) and (45) lead to:

$$\Delta t_{yup} = \left(\frac{Area_{up}}{Area_{down}}\right)\sqrt{\frac{2H}{g'}} = \theta_2\sqrt{\frac{2H}{g}} \tag{46}$$

The coefficient θ_2 accounts for the proportionality constant K of equation (44), the ratio between the mean areas of equation (40), and the use of the characteristic velocity. The power loss of the mean descending flow in the region of V_{down} is given by:

$$\dot{W}(V_{down}) = \rho g Q_R h_{loss} = K\rho Q_R \frac{V_{down}^2}{2} = \frac{K}{1+K}\rho g Q_R H \tag{47}$$

The rotating flow rate Q_R in the roller is the remaining parameter to be known. Because no variation of mass exists, it may be quantified at any position (or transversal section "S") **of the roller** (S_1, S_{up}, and S_{down}) as shown in Figure 8b. That is, Q_R=Q_1=Q_{up}=Q_{down}. Section S_1 is used here to quantify Q_R. A general equation for the velocity profile in section S_1 is used, following a power series of y, as:

$$\frac{V}{V_1} = \sum_{i=1}^{\infty}\beta_i\left(\frac{y}{\alpha H}\right)^i \tag{48}$$

αH expresses the maximum value of y in S_1, as shown in Figure 8b. β_i are coefficients of the power series, in which β_0=1. The mean velocity in section S_1 is obtained through integration, and Q_R is given by multiplying the mean velocity by the area αBH, furnishing, respectively:

$$\overline{V} = V_1 \sum_{i=1}^{\infty} \frac{\beta_i}{i+1}, \qquad Q_R = Q_1 = V_1 BH \sum_{i=1}^{\infty} \frac{\alpha\beta_i}{i+1} \tag{49}$$

Combining equations (47) and (49) results in:

$$\dot{W}\left(V_{down}\right) = \rho g V_1 B H^2 \theta_3, \qquad \theta_3 = \frac{K}{1+K} \sum_{i=1}^{\infty} \frac{\alpha\beta_i}{i+1} \tag{50}$$

Following similar steps for the flow in the region of V_1, the power consumption is given by:

$$h_{loss} = K \frac{V_1^2}{2g}, \qquad \dot{W}\left(V_1\right) = \rho \frac{V_1^3}{2} BH\theta_4, \qquad \theta_4 = K \sum_{i=1}^{\infty} \frac{\alpha\beta_i}{i+1} \tag{51}$$

Finally, repeating the procedures for the flow in the region of V_{up}, the power consumption is given by:

$$h_{loss} = K \frac{1}{2g} \left(\frac{H}{\Delta t_y} \right)_{up}^2 = \frac{K}{4\theta_2^2} H, \qquad \dot{W}\left(V_{up}\right) = \rho g V_1 B H^2 \theta_5, \qquad \theta_5 = \frac{K}{4\theta_2^2} \sum_{i=1}^{\infty} \frac{\alpha\beta_i}{i+1} \tag{52}$$

iii. Final equation for CV1 using the results of the global and intrinsic analyses

From equations (38), (41), and (50 to 52), it follows that:

$$g\, L_r \left(\frac{H}{\Delta t_y} \right)_{up} = \left(\frac{2\theta_3 + 2\theta_5 - 1}{\theta_1} \right) g V_1 H + \frac{\theta_4}{\theta_1} V_1^3 \tag{53}$$

Equations (46) and (53) finally produce:

$$g^{3/2} H^{1/2} L_r = \theta_6 g V_1 H + \theta_7 V_1^3, \quad \text{with} \quad \theta_6 = \sqrt{2}\,\theta_2 \left(\frac{2\theta_3 + 2\theta_5 - 1}{\theta_1} \right) \quad \text{and} \quad \theta_7 = \frac{\sqrt{2}\,\theta_2\theta_4}{\theta_1} \tag{54}$$

Nondimensional forms for the length L_r obtained from equations (54) are furnished below.

$$\frac{L_r}{H} = \theta_6 Fr_1 \sqrt{h^*} + \theta_7 \left(Fr_1 \sqrt{h^*} \right)^3 \tag{55}$$

$$\frac{L_r}{H} = \theta_6 Fr_H + \theta_7 Fr_H^3 \tag{56}$$

$$\frac{L_r}{h_1} = \theta_6 \frac{Fr_1}{\sqrt{h^*}} + \theta_7 \left(Fr_1\right)^3 \sqrt{h^*} \tag{57}$$

$$\frac{L_r}{h_2} = \theta_6 \frac{Fr_1}{h^\# \sqrt{h^*}} + \theta_7 \frac{Fr_1^3 \sqrt{h^*}}{h^\#} \tag{58}$$

In equations (55-58), $Fr_H = V_1/(gH)^{1/2}$, and $h^* = h_1/H$. As can be seen, the set of coefficients generated during the global and intrinsic analyses were reduced to only two, θ_6 and θ_7, which must be adjusted using experimental data. The normalizations (57) and (58) showed to be the most adequate for the experimental data analysed here.

2.1.2. Equation for sequent depths and Froude number using control volume 2 (CV2)

It remains to obtain an equation for the sequent depths. CV2 is now used in the stationary flow under study, as shown in Figure 9.

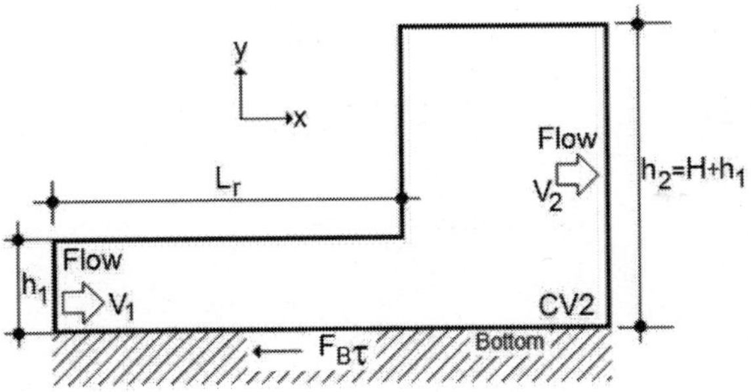

Figure 9. Control volume CV2 used to obtain an equation for the sequent depths.

The forces on CV2 in the contact surfaces between CV1 and CV2 vanish mutually (equilibrium of CV1), thus they are not indicated in Figure 9. In this case, two situations may be considered:

i. The force at the bottom ($F_{B\tau}$) is neglected.

ii. The force at the bottom is relevant.

The solutions in the sequence consider both cases.

i. The force at the bottom is neglected

From the mass conservation equation it follows that:

$$V_2 = V_1 / h^{\#} \tag{59}$$

From the momentum equation it follows that:

$$F_{B_\tau} = \rho B\left(h_2 - h_1\right)[V_1^2 / h^{\#} - g\left(h_1 + h_2\right)/2] \tag{60}$$

Assuming $F_{B\tau}$=0, equation (7) is reproduced, that is:

$$h^{\#} = \left(\sqrt{1+8Fr_1^2} - 1\right)/2 \tag{61}$$

Using h_1 as reference depth (for normalization procedures) equations (57) and (61) form the set of equations that relate the "sizes" h_1, h_2, and L_r in this study. Because $F_{B\tau}$=0, dissipation concentrates in the roller region (CV1). It is of course known that the forces at the interface between both volumes dissipate energy in CV2, but because the momentum equation already furnishes a prediction for h_2/h_1, it is not necessary to use the energy equation for this purpose. On the other hand, when considering the resistive force at the bottom, power losses must be considered in order to obtain predictions for $h^{\#}$.

ii. The force at the bottom is relevant

In this case, from the energy equation it follows that:

$$-\dot{W} = \rho B V_1 h_1\left(h_2 - h_1\right)\left(g - \frac{V_1^2}{h_2^2}\frac{\left(h_1 + h_2\right)}{2}\right) \tag{62}$$

The left side of equation (62) must consider the power transferred to CV1, denoted by $\dot{W}_{VC1}$, and the power lost due to the shear with bottom of the flow, denoted by $\dot{W}_{FB\tau}$. In mathematical form:

$$\dot{W} = \dot{W}_{VC1} + \dot{W}_{FB\tau} \tag{63}$$

From equations (41), (50), (51) and (52) it follows that:

$$\dot{W}_{VC1} = \rho B V_1\left(h_2 - h_1\right)[g\left(h_2 - h_1\right)\left(\theta_3 + \theta_5\right) + V_1^2\theta_4 / 2] \tag{64}$$

The velocity close to the bottom is ideally V_1. However, because of the expansion of the flow, it does not hold for the whole distance along the bottom. To better consider this variation, the power loss at the bottom was then quantified using a coefficient θ_8 for V_1, in the form:

http://dx.doi.org/10.5772/58963

$$\dot{W}_{FB\tau} = \theta_8 V_1 F_{B\tau} = \theta_8 V_1 \left\{ \rho B \left(h_2 - h_1 \right) [V_1^2 / h^\# - g\left(h_1 + h_2 \right) / 2] \right\} \tag{65}$$

Equations (61) to (65) produce the cubic equation:

$$\left[2\left(\theta_3 + \theta_5 \right) - \theta_8 \right] h^{\#3} + \left[2 - \theta_8 - 2\left(\theta_3 + \theta_5 \right) + \theta_4 Fr_1^2 \right] h^{\#2} - Fr_1^2 \left(1 - 2\theta_8 \right) h^\# - Fr_1^2 = 0 \tag{66}$$

The solution of the cubic equation (66) furnishes $h^\#$ as a function of Fr_1. Equations (57) and (66) form now the set of equations that relate h_1, h_2, Fr_1, and L_r for power dissipation occurring also in CV2. In section 3 of this chapter, equation (66) **is reduced to a second order equation**, based on the constancy of one of its solutions. Experimental results are then used to test the proposed equations.

The use of the power dissipation introduced coefficients in the obtained equations that affect all parcels having powers of h^* and $h^\#$. In order to adjust numerical values to θ_4, θ_8 and $2(\theta_3+\theta_5)$, the following form of equation (66) can be used in a multilinear analysis:

$$Fr_1^2 \left(1 + h^\# \right) - 2h^{\#2} = \theta_8 \left[2Fr_1^2 h^\# - h^{\#2} \left(1 + h^\# \right) \right] + 2\left(\theta_3 + \theta_5 \right) h^{\#2} \left(h^\# - 1 \right) + Fr_1^2 h^{\#2} \theta_4 \tag{67}$$

Because multilinear analyses depend on the distribution of the measured values (deviations or errors), different arrangements of the multiplying factors (for example, equations (67) and (68)) may generate different numerical values for θ_4, θ_8 and $2(\theta_3+\theta_5)$.

$$\frac{Fr_1^2 \left(1 + h^\# \right) - 2h^{\#2}}{Fr_1^2 h^{\#2}} = \theta_8 \left[\frac{2Fr_1^2 h^\# - h^{\#2} \left(1 + h^\# \right)}{Fr_1^2 h^{\#2}} \right] + 2\left(\theta_3 + \theta_5 \right) \frac{h^{\#2} \left(h^\# - 1 \right)}{Fr_1^2 h^{\#2}} + \theta_4 \tag{68}$$

2.2. Criteria for the function $h(x)$

2.2.1. The "depth deficit" criterion

The focus of the present section is the form of the surface (its profile). Some general characteristics of classical hydraulic jumps were considered, in order to develop a simple criterion, as follows:

i. The mean water depth tends to h_2 downstream from the toe of the jump.

ii. The mean slope of the surface is greater for greater "depth deficits", or differences between h_2 and the local mean water depth h (no undulating surfaces are presently considered).

iii. At the toe, the depth is h_1.

Using x as the longitudinal direction pointing downstream and with origin at the toe of the jump, the second characteristic described above (ii) is expressed mathematically as:

$$\frac{dh}{dx}\alpha\left(h_2-h\right) \qquad \text{or} \qquad \frac{dh}{dx}=\theta_{11}\left(h_2-h\right) \tag{69}$$

θ_{11} is a proportionality coefficient with dimension of m^{-1}. The integration of equation (69) using characteristics i and iii as boundary conditions produces:

$$\frac{h-h_1}{h_2-h_1}=1-e^{-\theta_{11}x} \qquad \text{(we define } \Pi=\frac{h-h_1}{h_2-h_1} \text{ for further calculations)} \tag{70}$$

The surface tends asymptotically to h_2 following an exponential function. The proportionality coefficient must be obtained from experimental data, which is done in section 3. Equation (70) is adequate for surface evolutions without inflexion points.

2.2.2. The "air inflow" criterion

[28] and [29] studied the transition from "black water" to "white water" in spillways and proposed that, if the slope of the surface is produced by the air entrainment, the transfer of air to the water may be expressed in the form:

$$\dot{c}=Kq\frac{dh}{dx} \tag{71}$$

$\dot{c}$ [s^{-1}] is the air transfer rate (void generation), q [m^{2}s^{-1}] is the specific water flow rate, h [m] is the total depth of the air-water flow, x [m] is the longitudinal axis and K [m^{-2}] is a proportionality factor. Equation (71) states that higher depth gradients for the same flow rate imply more air entraining the water, the same occurring for a constant depth gradient and higher flow rates. [30] showed that this equation allows obtaining proper forms of the air-water interfaces in surface aeration (stepped spillways).

Considering hydraulic jumps and a 1D formulation, an "ideal" two-step "expansion" of the flow is followed: 1) the flow, initially at velocity V_1 and depth h_1, aerates accordingly equation (71), expanding to ~h_2 ("close" to h_2); 2) the flow loses the absorbed air bubbles, decelerating to V_2 and attaining h_2. The geometrical situation considered in this first analysis is shown in Figure 10.

The evolution of the surface in the first step is obtained using the mass conservation equation:

$$\frac{\partial C}{\partial t}+u\frac{\partial C}{\partial x}+v\frac{\partial C}{\partial y}+w\frac{\partial C}{\partial z}=\left[\frac{\partial}{\partial x}\left(D_T\frac{\partial C}{\partial x}\right)+\frac{\partial}{\partial y}\left(D_T\frac{\partial C}{\partial y}\right)+\frac{\partial}{\partial z}\left(D_T\frac{\partial C}{\partial z}\right)\right]+\dot{p} \tag{72}$$

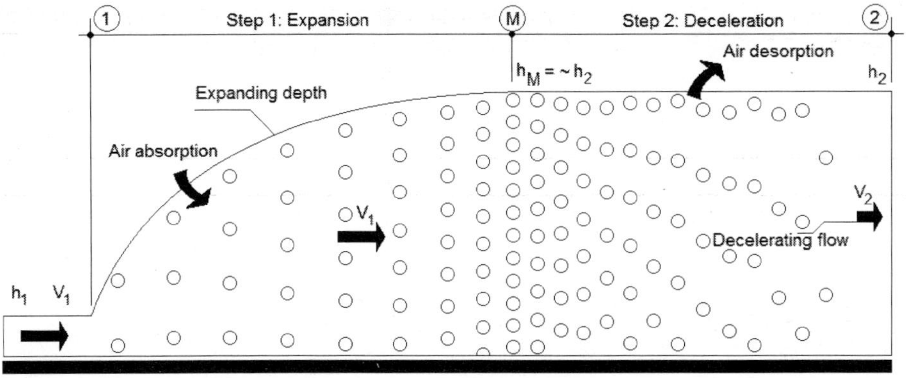

Figure 10. Geometrical condition of this analysis. 1D formulation. Please see also Figure 2.

C is the air concentration in the mixture. u, v, and w, are the velocities along x, y, and z, respectively. D_T is the turbulent diffusivity of the mixture (assumed constant), and $\dot{p}$ is the source/sink parcel. The stationary 1D equation simplifies to:

$$u\frac{\partial C}{\partial x} = D_T\frac{\partial^2 C}{\partial x^2} + \dot{p} \tag{73}$$

Equation (73) is normalized using the density of the air ρ_{air} (the void ratio ϕ is defined as ϕ=C/ρ_{air}), and the length of the jump, L_j, or roller, L_r. Thus we use here simply L (defining s=x/L). The result is:

$$\frac{\partial \phi}{\partial s} = \frac{D_T}{uL}\frac{\partial^2 C}{\partial s^2} + \dot{\phi} \tag{74}$$

By normalizing equation (71) to obtain ϕ, the mass discharge of the liquid, $\dot{m}$, was used, so that:

$$\phi = \frac{K\dot{m}}{L}\frac{dh}{ds} \tag{75}$$

K was adjusted to K'. In the 1D formulation the mixture is homogeneous in the cross section. The mean water content in the mixture is expressed as ϕ_{water}=h_1/h. h_1 is the supercritical depth and h=h_1+Z, being h the total depth and Z the length of the air column in the mixture. ρ_{air}<<ρ_{water}, so that:

$$\rho_1/\rho_M = h_M/h_1 , \qquad \phi = Z/h, \qquad h{=}h_1/(1-\phi) \tag{76}$$

The indexes 1 and M represent the two sections of the flow shown in Figure 10, and ρ is the water density. The integral mass conservation equation furnishes:

$$\dot{m} = \rho_1 h_1 V_1 = \rho_M h_M V_M \qquad \text{or} \qquad V_1 = V_M = u \tag{77}$$

From equations (74) to (77) it follows:

$$\frac{\partial \phi}{\partial s} = \frac{D_T}{V_1 L}\frac{\partial^2 \phi}{\partial s^2} + \frac{K'\dot{m}h_1}{L}\frac{\partial (1-\phi)^{-1}}{\partial s} \tag{78}$$

K' is a factor that may vary along the flow, and indicates the "facility" of the air absorption by the water. Considering arguments about diffusion of turbulence from the bottom to the surface, [30] proposed $K'=K''h^{-2}$. Equation (79) is then obtained:

$$\frac{\partial \phi}{\partial s} = \omega_1\frac{\partial^2 \phi}{\partial s^2} + \omega_2\frac{\partial \phi}{\partial s}, \qquad \text{where} \qquad \omega_1 = \frac{D_T}{V_1 L}, \qquad \omega_2 = \frac{K''\dot{m}h_1}{L} \tag{79}$$

The solution of equation (79), using boundary conditions 1) s=0, ϕ = 0, and 2) $s \to \infty$, $\phi \to \phi_\infty$, and defining IJ=(ω_2-1)/ω_1, is given by:

$$\phi = \phi_\infty\left(1-e^{-IJs}\right) \tag{80}$$

Equations (76) and (80) produce:

$$h/h_1 = 1/[1-\phi_\infty(1-e^{-IJs})] \tag{81}$$

In order to obtain a nondimensional depth like equation (70), a boundary layer analogy was followed assuming that at s=1 (length of the roller) we have $h/h_1=\beta h^\#$, where β is a constant "close" to the unity (as information, β=0.99 is the boundary layer value, but it does not imply its use here). For $s \to \infty$ we have $h/h_1=h^\#$. This implies that $\varphi_\infty=(h^\#-1)/h^\#$, so that the solution is given now by:

$$\Pi = \frac{h-h_1}{h_2-h_1} = \frac{(1-e^{-IJs})}{h^\# - (h^\#-1)(1-e^{-IJs})} \qquad IJ = -\ln\left[\frac{1-\beta}{\beta(h^\#-1)}\right] \tag{82}$$

Equation (82) is adequate for surface profiles with inflexion points.

2.3. The Random Square Waves rms criterion

The obtained mean profiles of the surface are useful to obtain the profile of the maximum fluctuations of the surface, which may be used in the design of the lateral walls that confine the hydraulic jump. In this sense, the method of Random Square Waves (RSW) for the analysis of random records is used here. Details of this method, applied to interfacial mass transfer, are found in [31, 32]. A more practical explanation is perhaps found in [33], where the authors present the rms function of random records, that is, the "equivalent" to the standard error function (RSW are bimodal records, but used to obtain approximate solutions for general random data).

Equations (70) and (82) express the normalized mean depth h along the hydraulic jump. The RSW approximation considers the maximum (H_2) and minimum (H_1) values of h, that is, it involves the local maxima of the fluctuations. Following the RSW method, the depth profile may be expressed as:

$$\frac{h-H_1}{H_2-H_1}=n \tag{83}$$

The normalized RSW rms function (a "measure" of the standard error) of the depth fluctuations is:

$$\frac{h'}{H_2-H_1}=\sqrt{n(1-n)}(1-\alpha) \tag{84}$$

h' is the rms value, and α is called "reduction function", usually dependent on x. From its definition it is known that $0\leq\alpha\leq1$ (see, for example, [33]). [31] and [32] present several analyses using a constant α value, a procedure also followed here as a first approximation. Undulating mean surface profiles, however, need a more detailed analysis.

Equations (70) and (82) to (84) allows obtaining equations for the height attained by fluctuations like waves and drops along the hydraulic jump, using procedures similar to those of normal distributions of random data. For normal distribution of data, 68.2%, 95.5% or 99.6% of the observed events may be computed by using respectively, 1, 2 or 3 standard errors in the prediction (for example). Figure 11 illustrates the comments, where μ represents the mean value and σ represents the standard error. The RSW method uses modified bimodal records (thus, the mentioned percentages may not apply), but the general idea of summing the mean depth and multiples of the rms value is still valid.

In order to relate n from equations (83) and (84) to Π from equations (70) and (82) it was considered that H_1 is zero (no fluctuation crosses the bottom), H_2 is given by $H_2=h_2+Nh_2'$(mean depth summed to N times the rms value), and $h_{wall}=h+Mh'$. Algebraic operations lead to:

$$\frac{h'}{h_2 - h_1} = (1-\alpha)\sqrt{\left(\Pi + \frac{1}{h^\# - 1}\right)\left(1 - \Pi + \frac{N^2 (1-\alpha)^2 h^\#}{h^\# - 1}\right)} \tag{85}$$

$$\frac{h_{wall}}{h_2} = \frac{1}{h^\#} + \left(1 - \frac{1}{h^\#}\right)\left\{\Pi + M(1-\alpha)\sqrt{\left(\Pi + \frac{1}{h^\# - 1}\right)\left(1 - \Pi + \frac{N^2 (1-\alpha)^2 h^\#}{h^\# - 1}\right)}\right\} \tag{86}$$

h_{wall} is used here to represent the mean height attained by the fluctuations. N and M are constants multiplied by the rms value ("replacing" the standard error) which originally expresses the percentage of cases assumed in the solution. In the present study (1-α) is taken as an adjustment coefficient. When analyzing the maximum value of h_2, the term between braces in equation (86) is substituted by J(max).

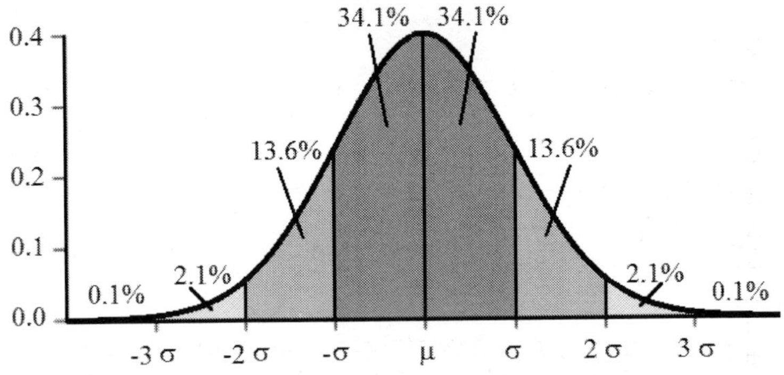

Figure 11. Normal curve showing the percentage of observed cases as the area limited by the curve.

3. Data analyses

3.1. Length of the roller

The data of [18] were first used to test equations (55-58). Different normalizations lead to different adjusted coefficients, because of the different distributions of errors. It was observed here that equation (57) shows the best adjustment to measured data, furnishing the coefficients shown in equation (87), and the adjustment of Figure 12. The obtained correlation coefficient was 0.996. L is used without index (r or j) because both lengths (roller and jet) are used in the following analyses.

$$\frac{L}{h_1} = 3.62\frac{Fr_1}{\sqrt{h^*}} - 0.157\left(Fr_1\right)^3\sqrt{h^*} \tag{87}$$

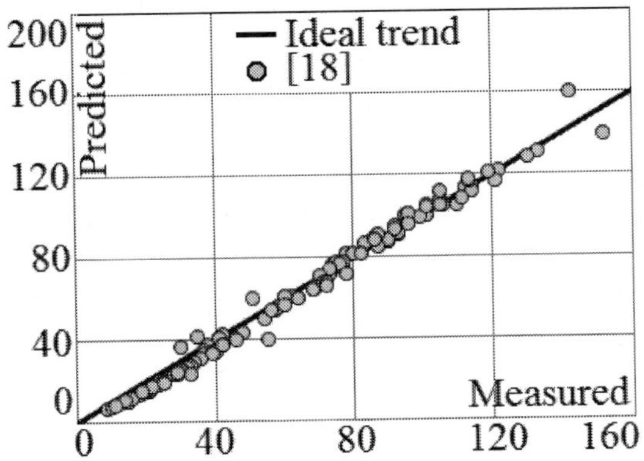

Figure 12. Adjustment of the data of [18] to equation (87).

Having obtained this first adjustment, other data of the literature were also compared to the formulation, obtaining the coefficients of Table 4 (R^2 are those of the multilinear regression). Figure 13 shows all predictions using equation (57) plotted against the measured value. The good linear correlation between predicted and measured values shows the adequacy of the present formulation. Further, different experimental arrangements and definitions of relevant parameters influence the coefficients θ_6 and θ_7, allowing the study of these influences. For the purposes of this study, which intends to expose the adequacy of the formulation, the literature data were used without distinguishing L_r and L_j, or the different ways used to obtain these lengths. Other influences, as the growing of the boundary layer at the supercritical flow, width of the channel, bottom roughness, bottom slope, distance to the downstream control (gate or step), use of sluice gates or spillways at the supercritical flow, among other characteristics, may also be studied in order to establish the dependence of θ_6 and θ_7 on these variables.

Table 4 shows that the coefficients depend on the experimental conditions. As an example, the data of [36] were obtained for several bed slopes. Table 4 presents the coefficients obtained for the slopes of 1° and 5°, suggesting variations in θ_6 and θ_7 (the component of the weight along the channel may influence, for example). Further, the data of [37] were obtained for rough beds, which coefficients may be compared to those for smooth beds of [38], suggesting that roughness also affects the coefficients ([38] also presents data for rough beds, allowing further comparisons). The data of [17] were obtained for flumes having as inlet structure a sluice gate or a spillway, which may be the cause of the different coefficients. The number of "points" of the different series is also shown, in order to allow verifying if the results depend on the "length" of the sample. Because coefficients are adjusted statistically, variations in θ_6 and θ_7 are expected when using only parts of a whole sample (like the two series of [18]). But also similar results were obtained for data of different authors, like [2], for 17.5°, and [17], for spillways, who furnished samples for 35 and 10 "points", respectively.

Author	θ_6	θ_7	R^2	Number of points
[18] Flume F	5.87	-0.586	0.991	22
[18] All data	3.62	-0.157	0.996	120
[34]	4.13	-0.157	0.995	30
[35]	5.10	-0.373	0.998	50
[2] 16.5°C	3,43	-0,191	0,999	17
[2] 17.5°C	2,75	-0,103	0,995	35
[36] 1° Bed slope	4,01	-0,177	0,9998	20
[36] 5° Bed slope	4,30	-0,339	0,9996	14
[37] Data for rough beds	2.76	-0.126	0.998	11
[38] Data for smooth beds	3.60	-0.139	0.994	20
[39]	2.81	-0.187	0.989	72
[15]	4.10	-0.312	0.999	6
[40]	4.65	-0.288	0.9994	31
[17] Spillways	2,79	-0.109	0.9999	10
[17] Sluice gates	2.94	-0.156	0.9999	8

Table 4. Values of θ_6 and θ_7 adjusted to data of different authors.

It is of course necessary to have a common definition of the lengths of the roller and the hydraulic jump (for the different studies) in order to allow a definitive comparison. Historically, different ways to obtain the lengths were used, which may add uncertainties to this first discussion of the coefficients. However, the good joint behavior of the different sets of data shown in Figure 13 emphasizes that such discussion is welcomed.

3.2. Sequent depths for the hypothesis "The force at the bottom is neglected"

Figure 3 already shows the good adjustment obtained using data of different authors. Different conditions of the channel may affect the predictions of equation (7) or (61), as shown in the literature, but the ratio between sequent depths, if working with the classical jump, is well predicted, as can be seen in Figure 3.

3.3. Sequent depths for the hypothesis "The force at the bottom is relevant"

The condition of "not-negligible bottom force" may be relevant, for example, for rough beds, which may affect the length of the jump and the subcritical depth. In order to illustrate this

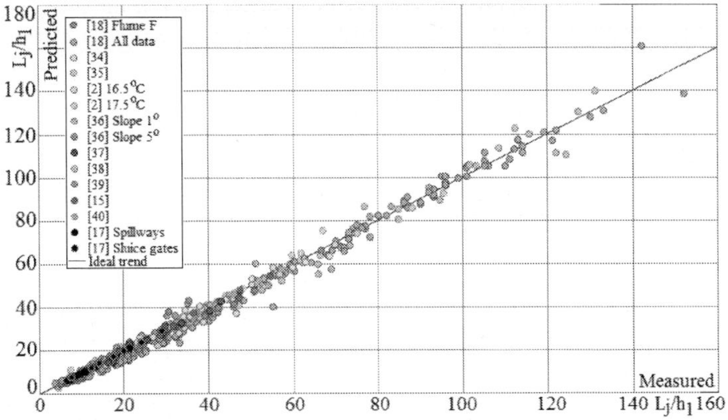

Figure 13. The predictions of L/h_1 for different experiments described in the literatures using the coefficients of Table 4 and equation (57).

condition, a multilinear regression analysis was made using equation (67) and the data of [37], furnishing θ_4=0.001229, θ_8=0.4227, and $2(\theta_3+\theta_5)$=0.5667. The determinant of equation (66) was negative for 10 experimental conditions. Thus, three real roots were obtained for these conditions. On the other hand, the determinant was positive for one experimental condition, leading to only one real root. When plotting the three calculated roots for $h^{\#}$ in relation to the measured values, Figure 14 was obtained (several points superpose).

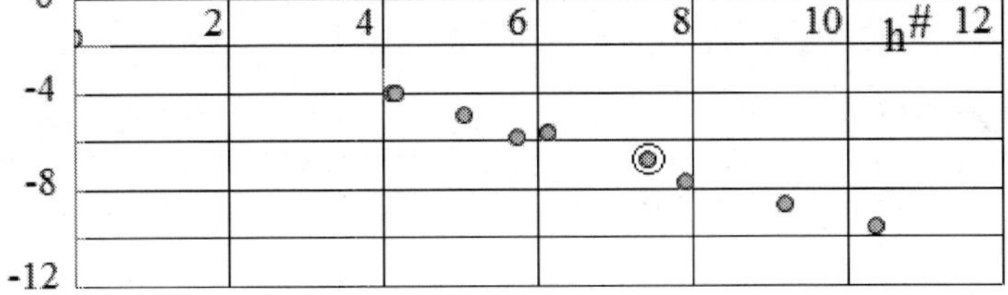

Figure 14. The three roots of $h^{\#}$ predicted with the cubic equation (66) for each experimental condition of [37]. The negative roots (in gray) are physically impossible. The positive roots (in white and black) show a constant depth ratio (Ghost Depth) and the reproduction of the observed values (inclined line). The gray point with a circle corresponds to the real root obtained for the positive determinant of equation (66). Several points superpose.

The constant $h^{\#}$ indicated by the horizontal line was obtained for the different sets of data analyzed here, assuming a proper value for each set. However, because it does not represent an observed result, it was named here "Ghost Depth", h_{GHOST}. From equation (66) this root is quantified as:

$$h_{\mathrm{GHOST}} = -\frac{2-\theta_8-2(\theta_3+\theta_5)}{2(\theta_3+\theta_5)-\theta_8} = \frac{(1-2\theta_8)\pm\sqrt{(1-2\theta_8)^2+4\theta_4}}{2\theta_4} \tag{88}$$

Considering the first equality of equation (88), and dividing the cubic equation (66) by the monomial obtained with this solution, following second degree equation is obtained:

$$h^{\#2} - 2\theta_9 Fr_1^2 h^{\#} - \theta_{10} Fr_1^2 = 0 \tag{89}$$

Where:

$$\theta_9 = -\frac{\theta_4}{2[2(\theta_3+\theta_5)-\theta_8]}, \qquad \theta_{10} = \left\{\frac{(1-2\theta_8)}{2(\theta_3+\theta_5)-\theta_8} + \frac{\theta_4[2-\theta_8-2(\theta_3+\theta_5)]}{[2(\theta_3+\theta_5)-\theta_8]^2}\right\}$$

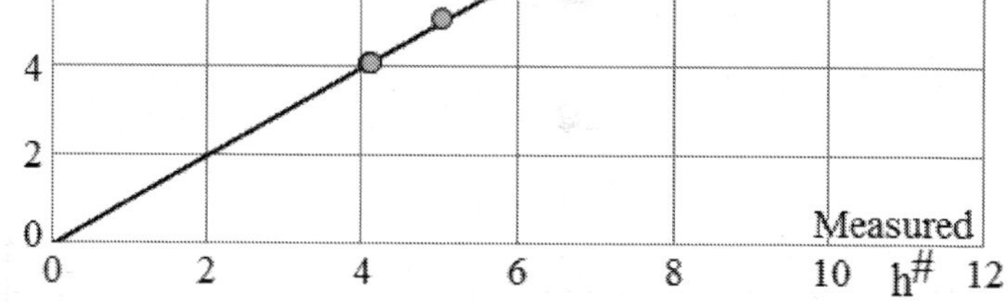

Figure 15. Calculated values of $h^{\#}$ obtained with equation (90) and the data of [37]. The number of experimental conditions is 11, but several points superpose.

As can be seen, also in this case the final equation is a second order equation and depends on only two coefficients, θ_9 and θ_{10}. The solution for $h^{\#}$ is given by:

$$h^{\#} = \theta_9 Fr_1^2 \pm \sqrt{\left[\theta_9 Fr_1^2\right]^2 + \theta_{10} Fr_1^2} \tag{90}$$

The adjusted coefficients produced the graph of Figure 15 for the positive root of equation (90) and the data of [37].

The good result of Figure 15 induced to analyze more data of the literature using equations (89) and (90). Table 5 shows the coefficients obtained from multilinear regression analyses applied to the original data, and Figure 16 presents all data plotted together. The values of R^2 in Table 5 are those obtained in the multilinear regression.

Author	θ_9	θ_{10}	$\sqrt{\theta_{10}}$	R^2	Number of points
[18] Flume F	0.0158	1.44	1.20	0.9993	22
[18] All data	0.00908	1.62	1.27	0.98	120
[34]	0.0138	1.51	1.23	0.998	30
[35]	0.0318	1.27	1.13	0.998	50
[2] 16.5°C	0.00630	1.50	1.22	0,9998	17
[2] 17.5°C	-0.00588	1.78	1.33	0,999	35
[36] 1° Bed slope	0.0437	0.786	0.887	0,9998	20
[36] 5° Bed slope	0.0561	1.14	1.07	0,9996	14
[37] Data for rough beds	0.00375	0.982	0.991	0.998	11
[41] Data for smooth beds	-0.00800	2.52	1.59	0.98	20
[38] Data for smooth beds	0.0165	1.14	1.07	0.990	20
[39]	0.0151	1.50	1.22	0.9993	72
[15]	0.00265	1.72	1.31	0.99995	6
[42] Data for rough beds	-0.00634	1.20	1.10	0.9994	8
[43]	0.0122	1.28	1.13	0.998	10
[40]	0.0684	0.524	0.724	0.98	31
[17] Spillways	0.0119	1.78	1.33	0.993	10
[17] Sluice gates	-0.0320	2.36	1.54	0.96	8

Table 5. θ_9 and θ_{10} (and $\sqrt{\theta_{10}}$) adjusted to data of different authors.

The coefficients θ_9 and θ_{10} of Table 5 show that, for most of the cases, $\theta_{10} >> \theta_9$, which implies $h^{\#} = \theta_9 Fr_1^2 + \sqrt{\theta_{10}} Fr_1$ for the usual range of Fr_1. In many cases, the quadratic term of Fr_1 is still much smaller than the linear term. For example, the data of [15], obtained for $4.38 \leq Fr_1 \leq 9.26$ produce $\theta_9 Fr_1^2 = 0.227$ and $\sqrt{\theta_{10}} Fr_1 = 12.1$ as their highest values. This fact suggests to use simpler forms of equation (90) like $h^{\#}=\sqrt{\theta_{10}}Fr_1+Constant$ or $h^{\#}=\sqrt{\theta_{10}}Fr_1$. Both forms appear in the literature. The mean value of $\sqrt{\theta_{10}}$ in table 5 is 1.19, showing that the coefficients of Fr_1 stay around the unity. [44], for example, used $h^{\#}=1.047Fr_1+0.5902$, while [37] simply suggested $h^{\#}=Fr_1$ for their data. Equation (8) of the present chapter is a further example.

A very positive aspect of equation (90) is that it is based on the physical principles of conservation of mass, momentum and energy. The simpler forms of the literature follow directly when analyzing the magnitude of the different parcels of the final equation. That is, the simpler forms are in agreement with the conservation principles (being not only convenient results of dimensional analyses).

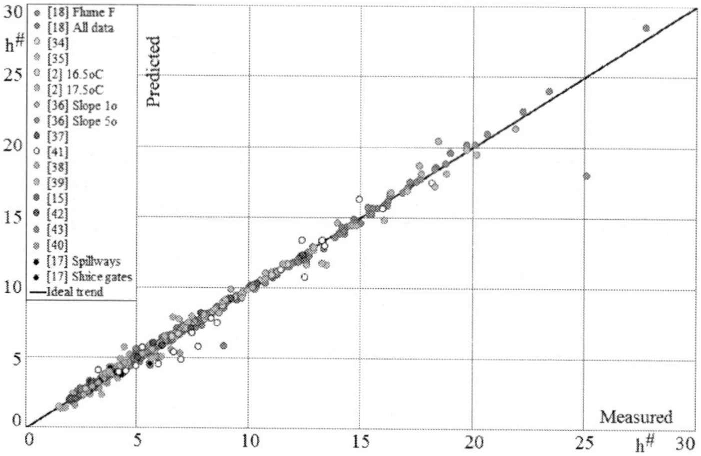

Figure 16. The predictions of h_2/h_1 for different experiments described in the literatures using the coefficients of Table 5 and equation (90).

As an additional information, the coefficient of Fr_1 in the simpler forms tends to be lower than $\sqrt{2}$, the value derived from equations (7) or (61) for no shear force at the bottom of the jump.

From the data analysis performed here, and the conclusions that derive from the presence of the Ghost Depth in the cubic equation, the equations for the geometrical characteristics of hydraulic jumps of case ii of section 2.1.2 (The force at the bottom is relevant) simplify to

equation (57) $\frac{L_r}{h_1}=\theta_6\frac{Fr_1}{\sqrt{h^*}}+\theta_7(Fr_1)^3\sqrt{h^*}$ and equation (90) $h^{\#}=\theta_9 Fr_1^2\pm\sqrt{[\theta_9 Fr_1^2]^2+\theta_{10}Fr_1^2}$, equations already presented in the text.

Considering the deduction procedures followed in this study, the values of θ_6, θ_7, θ_9 and θ_{10} may depend on the factors that determine the resistance forces (shear forces), similarly to the friction factor in uniform flows. Additionally, because $h^*=(h^{\#}-1)^{-1}$, the nondimensional length of the roller may be written as a function only of Fr_1 and the different coefficients.

3.4. Water depth profiles

3.4.1. Profiles without inflexion points

The data of [17] and [25], obtained for spillways, are shown in Figure 17. It is possible to admit no inflexion points in this set of surface profiles. So, equation (70) was used to quantify the surface evolution, furnishing equation (91). The adjustment produced θ_{11}=2.9, implying that x/L=1 (the roller length) is attained for Π=0.95, that is, for 95% of the final depth difference h_2-h_1. Such conditional definition of the length is similar to that used for boundary layer thicknesses, and seems to apply for hydraulic jumps.

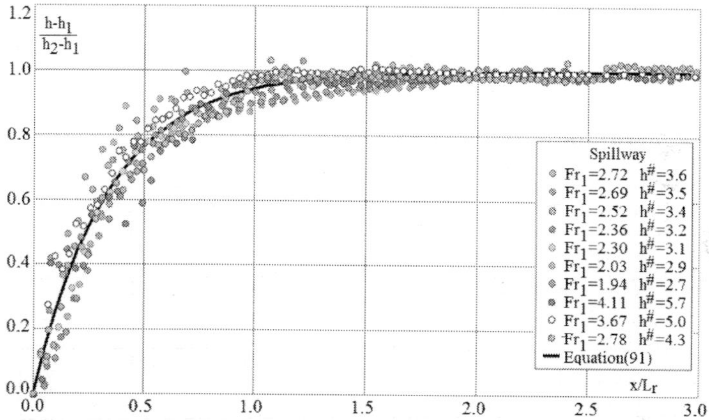

Figure 17. Normalized measured profiles obtained by [17] and [25], adjusted to the exponential prediction (equation 91), for profiles without inflexion points.

In the present analysis, the normalized profile assumes the form

$$\frac{h-h_1}{h_2-h_1}=1-e^{-2.9x/L} \tag{91}$$

3.4.2. Profiles with inflexion points

Equation (82) may be used with any option of $h^{\#}$, that is, equation (61) or (90), or the simpler form $h^{\#}=\sqrt{\theta_{10}}+Constant$. As example, equation (61) is used, generating equation (92) for β=0.99. For comparison, the simpler form of $h^{\#}$ is used (as proposed by [37]) with $\sqrt{\theta_{10}}$=1, $Constant$=0, and β=0.99, resulting $h^{\#}=Fr_1$ and equation (93). Figure 18a shows the comparison between equation (92) and the envelope of the measured data of [37]. Figure 18b shows the comparison between equations (92) and (93) for Fr_1=7. Both equations show the same form, with the results of equation (93) being somewhat greater than those of equation (92).

$$\frac{h-h_1}{h_2-h_1}=\frac{2(1-e^{-IJs})}{(\sqrt{8Fr_1^2+1}-1)-(\sqrt{8Fr_1^2+1}-3)(1-e^{-IJs})} \qquad IJ=-\ln\left[\frac{1}{99(\sqrt{8Fr_1^2+1}-3)}\right] \tag{92}$$

$$\frac{h-h_1}{h_2-h_1}=\frac{(1-e^{-IJs})}{Fr_1-(Fr_1-1)(1-e^{-IJs})} \qquad IJ=-\ln\left[\frac{1}{99(Fr_1-1)}\right] \tag{93}$$

Figure 19 was obtained for the data of [17] using equation (92), but with β=0.97, that is, with the exponent IJ given by:

$$IJ=-\ln\{6/[97(\sqrt{8Fr_1^2+1}-3)]\} \tag{94}$$

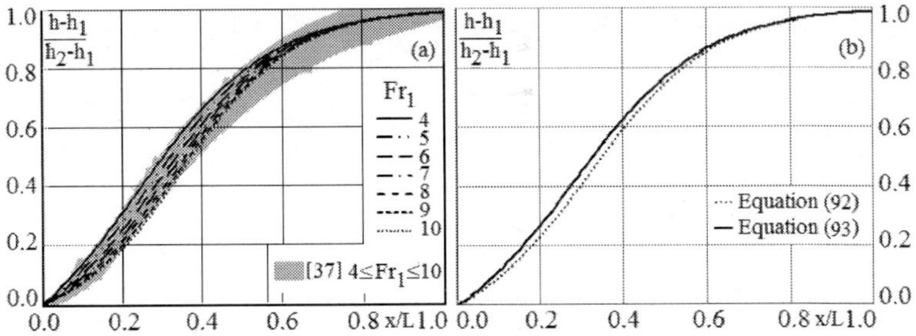

Figure 18. a) The lines are the predictions of equation (92) for the values of Fr_1 shown in the graph (β=0.99). The gray region is the envelope of the data of [37]; b) Comparison between equations (92) and (93) for events with and without bed shear force, and Fr_1=7.0.

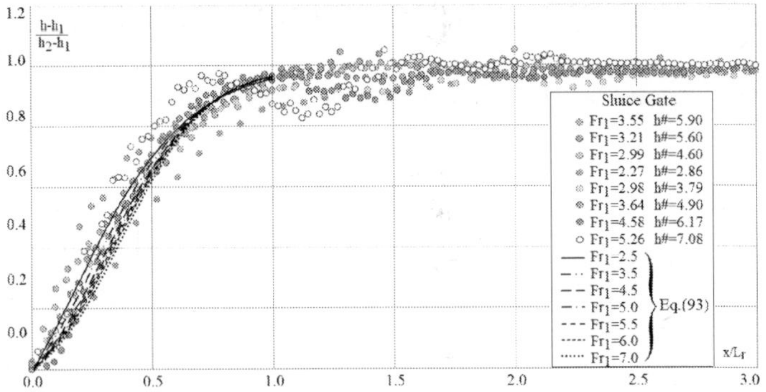

Figure 19. Normalized measured profiles obtained by [17] and [25], and predictions of equation (93) for profiles with inflexion points (β=0.97).

3.4.3. Further normalizations

[21] presented the experimental water profiles shown in Figure 20b. As can be seen, using their normalization the curves intercept each other for the experimental range of Fr_1. In order to verify if the present formulation leads to a similar behavior, a first approximation was made using equation (93) for β=0.95, that is, with the exponent IJ given by

$$IJ = -\ln\{5/[95(Fr_1 - 1)]\}$$

x/L was transformed into $x/(h_2-h_1)$ by multiplying $[x/(h_2-h_1)][h_1/L][h^{\#}-1]$. The coefficients θ_6 and θ_7 of Table 4 for the data of [37] were used to calculate L/h_1, together with $h^{\#}=Fr_1$. The calculated curves shown in Figure 20a also intercept each other, obeying similar relative positions when compared to the experimental curves. The results of the formulation thus follow the general experimental behavior.

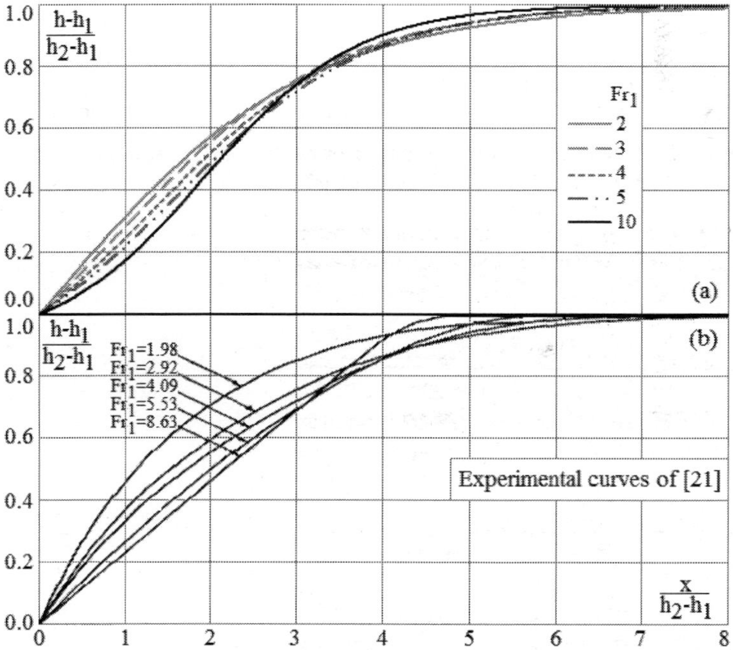

Figure 20. Comparison between observed and calculated surface profiles; a) Calculated curves; b) Experimental curves.

Various values of β were used in the present section (3.4), in order to obtain good adjustments. In each figure (18, 19, and 20) β was maintained constant for the family of curves, but it may not be always the case, considering previous results of the literature. For example, [45] suggested that the ratio between the final depth of the roller and h_2 is a function or Fr_1. Thus, each curve may have its own β.

3.5. Depth fluctuations and total depths (wall heights)

3.5.1. Predictions based on empirical formulation

As mentioned in item 2.3, oscillations of the free surface may be relevant for the design of side walls. [46] recommended, for the height of the wall, h_{wall}=1.25h_2. [22] performed measurements of water depths and fluctuations along a hydraulic jump in a 40 cm wide horizontal rectangular channel, and imposing Fr_1=3.0. The measurements were made using ultrasonic acoustic sensors. The mean ratio h/h_2 attained a maximum value in the roller region, h_{max}/h_2=1.115. The maximum normalized standard error of the fluctuations along the hydraulic jump was h'_{max}/h_1=0.32 for this condition. [47], based on data of several sources, showed that standard deviation of the fluctuations of the depths in hydraulic jumps increase with the Froude number, suggesting the empirical equation (94), for Fr_1 between 1.98 and 8.5.

$$h'_{max}/h_1 = 0.116(Fr_1 - 1)^{1.235} \tag{95}$$

This equation furnishes, for the condition Fr_1=3.0 studied by [22], the prediction h'_{max}/h_1=0.27, about 16% lower of the experimental value, suggesting further studies. In the present section the height h_{wall} is quantified following different approximations, and using experimental data. A first evaluation of h_{wall} may be given by the sum of the maximum depth and a multiple of the standard deviation (multiple given by the factor N), that is:

$$h_{wall}/h_2 = h_{max}/h_2 + N\,h'_{max}/(h_1 h^{\#}) \tag{96}$$

Using equations (95) and (96), no bed shear force, that is $h^{\#}=[(1+8Fr_1^2-1)^{1/2}-1]/2$, and correcting the predicted value of 0.27 to the observed value of 0.32, we have:

$$\frac{h_{wall}}{h_2} = 1.115 + \frac{0.275(Fr_1 - 1)^{1,235}}{\left(\sqrt{1+8Fr_1^2} - 1\right)}, \qquad (N=1) \tag{97}$$

$$\frac{h_{wall}}{h_2} = 1.115 + \frac{0.550(Fr_1 - 1)^{1,235}}{\left(\sqrt{1+8Fr_1^2} - 1\right)}, \qquad (N=2) \tag{98}$$

$$\frac{h_{wall}}{h_2} = 1.115 + \frac{0.825(Fr_1 - 1)^{1,235}}{\left(\sqrt{1+8Fr_1^2} - 1\right)}, \qquad (N=3) \tag{99}$$

As an example, Figure 21 presents different images of surface elevations due to waves, which illustrates a depth fluctuation. Figure 21a presents a sequence of images of a wave passing the hydraulic jump, and Figure 21b shows the relative depth attained by this wave. The approximate value h_{max}/h_2=1.40 is compared with equations (97-99).

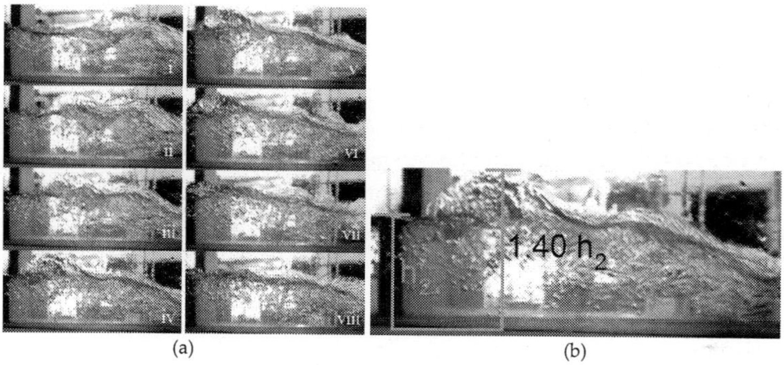

Figure 21. a) Oscillations of the free surface due to a wave in a hydraulic jump for Fr_1=3.5, b) Relative depth attained by the wave shown in Figure 21a iv.

Equations (97, 98, 99) use 1, 2 and 3 standard errors (68.2%, 95.5% and 99.6% of the observed events, respectively, see Figure 13). Figure 22 presents the curves of the three equations, showing that the additional height observed in Figure 21b is better represented by equation (99), for 99.6% of the cases. Equation (97) furnishes values closer to the suggestion of [46] (h_{wall}/h_2=1.25).

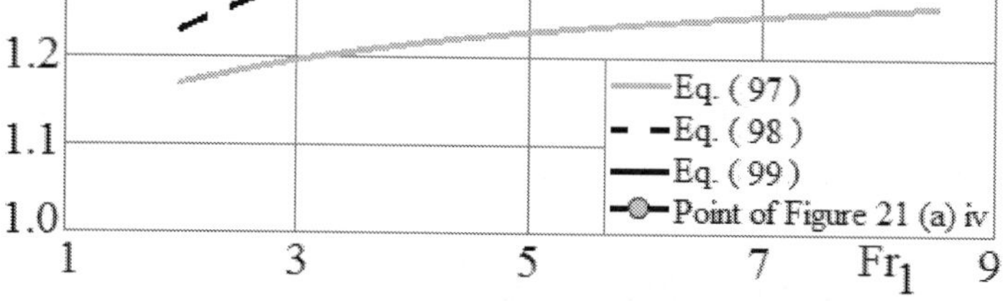

Figure 22. Equations (97, 98, 99), for the maximum depth as a function of the Froude number.

3.5.2. Predictions based on theoretical reasoning

Equations (85, 86) and (91) are used, with N=0.8, M=3.3 and α=0.82. h' and h_{wall} are given by:

$$\frac{h'}{h_2-h_1}=0.18\sqrt{\left(1-e^{-2.9x/L}+\frac{1}{h^{\#}-1}\right)\left(e^{-2.9x/L}+\frac{0.0207h^{\#}}{h^{\#}-1}\right)} \tag{100}$$

$$\frac{h_{wall}}{h_2}=1-\left(\frac{h^{\#}-1}{h^{\#}}\right)e^{-2.9x/L}+0.594\sqrt{\left[1-\left(\frac{h^{\#}-1}{h^{\#}}\right)e^{-2.9x/L}\right]\left[\left(\frac{h^{\#}-1}{h^{\#}}\right)e^{-2.9x/L}+0.0207\right]} \tag{101}$$

Predictions of equation (101) were compared with depth data obtained by [17]. The comparison is shown in Figure 23. The normalized mean depth of the data increases continuously with x/L_r, but when added to three times the standard error, the points show a very slight maximum. For the predictive calculations, it was assumed H_1=0 (a limiting value), but any value $0 \le H_1 \le h_1$ is possible. The value M=3.3 multiplied by the rms function of the RSW method is close to the "three standard error" used for h_{wall}/h_2. For Fr_1=2.36 the maximum calculated value is h_{wall}/h_2=1.11, lower than the suggested value of [46], that is, h_{wall}/h_2=1.25. Figure 24 shows the data measured by [17], where the vertical lines are the measured points.

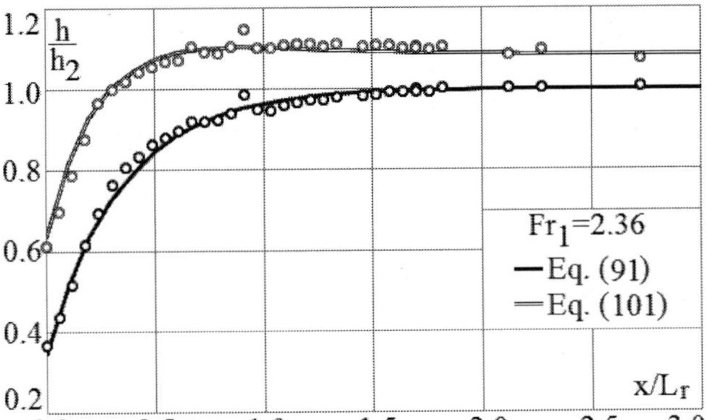

Figure 23. Measured surface profile (lower "o"), and 3 standard errors added to it (upper "o"). Data of [17]. Calculated profile (equation 91) and 3.3 times the rms value of the RWS method added to it (equation 101).

Predictions similar to those presented in Figure 22 need only the maximum value of the function along x/L_r. Indicating this value as "J(max)", equations (101) and (61) furnish, for no bed shear force:

$$\frac{h_{wall}}{h_2} = \frac{2}{\sqrt{1+8Fr_1^2}-1} + \left(\frac{\sqrt{1+8Fr_1^2}-3}{\sqrt{1+8Fr_1^2}-1}\right) J(\max_1), \qquad \text{adjusted } J(\max_1)\text{=}1.515 \tag{102}$$

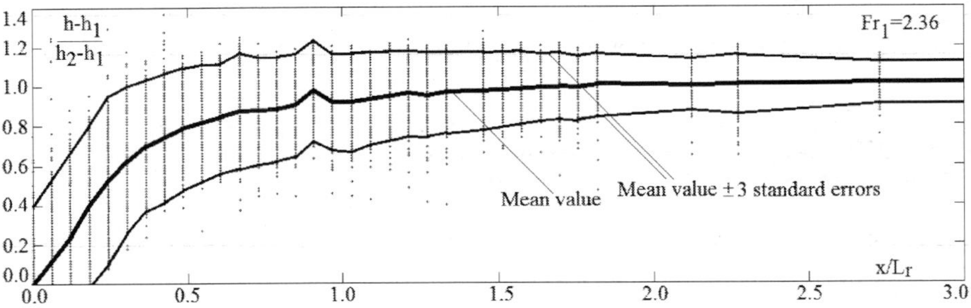

Figure 24. Measurements of [17] used in Figure 23.

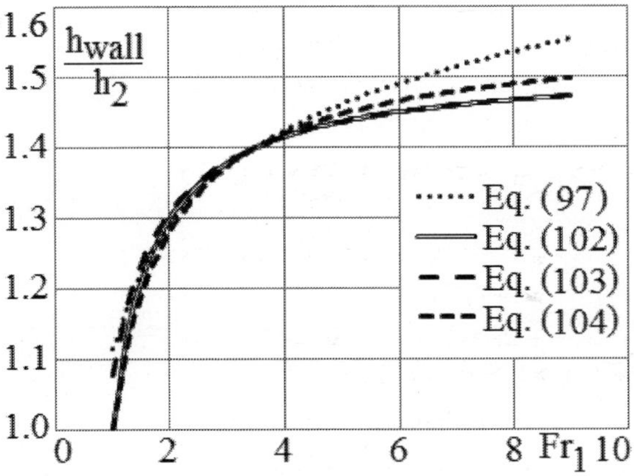

Figure 25. Equations (97), (102), (103) and (104) for the maximum depth as a function of the Froude number. J(max) adjusted for h_{wall}/h_2=1.4 and Fr_1=3.5.

For bed shear forces the form $h^{\#}=\sqrt{\theta}\ {}_{10}Fr_1+Constant$ can be used. As example, equation (101) together with equation (8) or $h^{\#}=Fr_1$ (see [37]) furnish, respectively:

$$\frac{h_{wall}}{h_2} = \frac{1}{1.29Fr_1-0.116} + \left(\frac{1.29Fr_1-1.116}{1.29Fr_1-0.116}\right) J(\max_2), \qquad \text{adjusted } J(\max_2)\text{=}1.518 \tag{103}$$

$$\frac{h_{wall}}{h_2} = \frac{1}{Fr_1} + \left(\frac{Fr_1 - 1}{Fr_1}\right) J(\max_3), \qquad \text{adjusted } J(\max_3)\text{=1.560} \tag{104}$$

The theoretical predictions of equations (99), (102), (103) and (104) on the Froude number (Fr_1) are shown in Figure 25. J(max) was adjusted in each equation to furnish h_{wall}/h_2=1.4 for Fr_1=3.5. J(max) also corresponds to the maximum value of h_{max}/h_2 for each prediction.

The theoretical predictions of h_{wall}/h_2 follow the general trend obtained when using the empirical approximation of [47], that is, higher h_{wall}/h_2 are obtained for higher Fr_1 (for the observed Fr_1 range). However, the condition h_{wall}/h_2=1.4 for Fr_1=3.5 was observed only in laboratory scale (see Figure 21). Scale-up studies are thus needed.

4. Conclusions

The conclusions are presented as suggestions of use for conditions similar to those of the mentioned experimental studies. The numbers of the equations refer to their location in the text.

4.1. Basic lengths

Condition: No bed resistance (no bed shear forces).

For the length of the roller, use (see Table 4 and Figure 13):

(Applied principles: conservation of mass, momentum, and energy).

Equation (57): $\frac{L_r}{h_1} = \theta_6 \frac{Fr_1}{\sqrt{h^*}} + \theta_7 (Fr_1)^3 \sqrt{h^*}$

For the sequent depths, use (see Figure 3):

(Applied principles: conservation of mass and momentum).

Equation (61): $h^{\#} = \left(\sqrt{1 + 8Fr_1^2} - 1\right)/2$

Condition: Relevant bed resistance (presence of shear forces).

For the length of the roller, use (see Table 4 and Figure 13):

(Applied principles: conservation of mass, momentum, and energy).

Equation (57): $\frac{L_r}{h_1} = \theta_6 \frac{Fr_1}{\sqrt{h^*}} + \theta_7 (Fr_1)^3 \sqrt{h^*}$

For the sequent depths, use (see Table 5 and Figure 16):

(Applied principles: conservation of mass, momentum, and energy).

Equation (90): $h^{\#}=\theta_9 Fr_1^2+\sqrt{[\theta_9 Fr_1^2]^2+\theta_{10} Fr_1^2}$

As shown in the text (see Table 5), the magnitude of the parcels of equation (90) also allows using the simplified forms, linked to the three principles of conservation used in equation (90):

$$h^{\#}=\sqrt{\theta_{10}}Fr_1+\theta_9 Fr_1^2$$

$$h^{\#}=\sqrt{\theta_{10}}Fr_1+Constant$$

$$h^{\#}=\sqrt{\theta_{10}}Fr_1$$

4.2. Surface evolution (surface profile)

For surfaces without inflexion point, use (see Figure 17):

(Applied "principle": Depth deficit).

Equation (70): $\frac{h-h_1}{h_2-h_1}=1-e^{-\theta_{11}^{*}x/L}$ with $\theta_{11}^{*}=L\,\theta_{11}$

The data analyzed in the present study furnished $\theta_{11}^{*}\approx 2.9$. But the evolution of the surface depends on operation conditions, and more studies are necessary.

For surfaces with inflexion point, use (see Figures 18, 19, 20):

(Applied principles: Conservation of mass (air and water) and basic air transfer equation (void generation).

Equation (82): $\frac{h-h_1}{h_2-h_1}=\frac{(1-e^{-IJs})}{h^{\#}-(h^{\#}-1)(1-e^{-IJs})}$ with $IJ=-\ln\left[\frac{1-\beta}{\beta(h^{\#}-1)}\right]$

The exponential form is maintained in the equation for simplicity.

4.3. Fluctuations and related depths (for lateral walls)

(Applied principle: rms of fluctuations as obtained in the RSW method).

For the dependence with the Froude number Fr_1, use:

Equation (86): $\frac{h_{wall}}{h_2}=\frac{1}{h^{\#}}+\left(1-\frac{1}{h^{\#}}\right)J(\max)$

From laboratory observations for Fr_1=3.5 (see Figure 21), the value of $J(\max)$ in this study varied between 1.52 and 1.56. But scale-up procedures are still necessary. $h^{\#}$ depends on the situation under study: with or without bed shear forces (using the equations for $h^{\#}(Fr_1)$ presented here).

For the evolution of the height attained by fluctuations along the jump, use:

Equation (85): $\frac{h'}{h_2-h_1}=(1-\alpha)\sqrt{\left(\Pi+\frac{1}{h^{\#}-1}\right)\left(1-\Pi+\frac{N^2(1-\alpha)^2h^{\#}}{h^{\#}-1}\right)}$

Equation (86): $\frac{h_{wall}}{h_2}=\frac{1}{h^{\#}}+\left(1-\frac{1}{h^{\#}}\right)\left\{\Pi+M(1-\alpha)\sqrt{\left(\Pi+\frac{1}{h^{\#}-1}\right)\left(1-\Pi+\frac{N^2(1-\alpha)^2h^{\#}}{h^{\#}-1}\right)}\right\}$

h_{wall} is used here to represent the mean total height attained by the fluctuations. N and M are constants. In the present study N=0.8 (close to 1, for the equivalent to the standard error), M=3.3 (close to 3, considering more events) and α=0.8 ($0 \leq \alpha \leq 1$ from the RSW method). Π was represented for a situation without inflexion points, that is, using equation (70) or (91).

4.4. Challenges

Having a set of equations for geometrical aspects of hydraulic jumps, which fit observed characteristics of different experimental situations, following questions need attention:

- Table 4 shows a somewhat broad range of values for coefficients θ_6 and θ_7 (although each of them maintains a coherent magnitude). In this table no distinction was made between the lengths of the roller and the jump itself. Also the different definitions and measurement procedures were not distinguished. A "standard definition" for the roller length and the jump length would help to compare different experimental results. In addition, the influences of experimental conditions on the coefficients may be studied in a systematic way (geometry, size, distribution of roughness elements, slope of the bed, etc., like usual studies of resistance factors or coefficients), which certainly will help the practical use of equation (57).

- Table 5 shows that the coefficients θ_9 and θ_{10} also maintain coherent magnitudes. In this case, $\sqrt{\theta_{10}}$ stays around the unity for the experimental data used here (mean value of 1.19). Also here a systematic study may furnish the dependence of the coefficients on controlled parameters, like roughness and slope. Such study will help the practical use of equation (90) and its simpler forms.

- The form of the surface for the conditions "with inflexion point" and "without inflexion point" was considered here for monotonic growing profiles. The form of the profile is important to detect the end of strong surface variations. In this case, a conditional length may be defined, similarly to the boundary layer thickness. In the present study the values 95%, 97% and 99% of $(h-h_1)/(h_2-h_1)$ were used to define the "roller length" or the "jump length", but it was mentioned that the literature already shows proposals in this sense, which may imply in broader ranges of values. A "standard definition" in this sense, would probably diminish the "imprecisions" of the multiple definitions and measurement strategies found in the literature. Additionally, ultrasound probes and high speed cameras have shown to be adequate to obtain information of the surface evolution, enabling to verify such conditional definitions. Undulating surfaces were not considered in the present study.

- The use of the conclusion $h'/(H_2-H_1)=\sqrt{n(1-n)}(1-\alpha)$ of the RSW method shows that it is possible to have a first prediction of the form of the height attained by the fluctuations along the jump based on the form of the mean surface profile for monotonic growing surfaces. More studies could establish adequate values (or functions) for α for the different experimental conditions, for example.

http://dx.doi.org/10.5772/58963

Author details

Harry Edmar Schulz[1*], Juliana Dorn Nóbrega[1], André Luiz Andrade Simões[2], Henry Schulz[3] and Rodrigo de Melo Porto[1]

*Address all correspondence to: heschulz@sc.usp.br

1 University of São Paulo, Brazil

2 Federal University of Bahia, Brazil

3 Educational Cooperative – Educativa São Carlos, Brazil

References

[1] Elevatorski EA. Hydraulic Energy Dissipators. New York: McGraw-Hill; 1959.

[2] Hager WH, Bremen R, Kawagoshi N. Classical hydraulic jump: length of roller. Journal of Hydraulic Research 1990;28 591-608.

[3] Ortiz JP. Macroturbulência de escoamentos à jusante de estruturas de dissipação por ressalto: estudo teórico-experimental. MSc Thesis. University of São Paulo; 1982.

[4] Ortiz JP. Estrutura e resolução de pressões flutuantes - sua análise randômica na base do ressalto hidráulico. PhD Thesis. University of São Paulo; 1989.

[5] Gil MML. Ressalto hidráulico em pressão a jusante em conduto circular inclinado. MSc Thesis. University of São Paulo; 1991.

[6] Yamashiki Y. Modelação matemática e física da erosão em presença de ressalto hidráulico. MSc Thesis. University of São Paulo; 1994.

[7] Gomes JF. Visualização de escoamento macroturbulento - ressalto hidráulico a jusante de uma comporta. MSc Thesis. Federal University of Rio Grande do Sul; 2000.

[8] Trierweiler Neto EF. Avaliação do campo de pressões em ressalto hidráulico formado a jusante de uma comporta com diferentes graus de submergência. MSc Thesis. Federal University of Rio Grande do Sul; 2006.

[9] Wiest RA. Avaliação do campo de pressões em ressalto hidráulico formado a jusante de um vertedouro com diferentes graus de submergência. MSc Thesis. Federal University of Rio Grande do Sul; 2008.

[10] Alves AAM. Caracterização das solicitações hidrodinâmicas em bacias de dissipação por ressalto hidráulico com baixo número de Froude. MSc Thesis. Federal University of Rio Grande do Sul; 2008.

[11] Quevedo DM. Análise de pressões junto ao fundo no ressalto hidráulico formado a jusante de um vertedouro através da distribuição bivariada de valores extremos. PhD Thesis. Federal University of Rio Grande do Sul; 2008.

[12] Cerezer SM. Uso da teoria de valores extremos para estimar valores de pressões hidrodinâmicas em um ressalto hidráulico formado a jusante de um vertedouro: o caso da UHE Porto Colômbia. PhD Thesis. Federal University of Rio Grande do Sul; 2008.

[13] Teixeira ED. Efeito de escala na previsão dos valores extremos de pressão junto ao fundo em bacias de dissipação por ressalto hidráulico. PhD Thesis. Federal University of Rio Grande do Sul; 2008.

[14] Ortiz JP. Fundamentos da turbulência de escoamentos incompressíveis aplicados à engenharia. Associate Professor Thesis. University of São Paulo; 2011.

[15] Dai Prá M. Uma abordagem para determinação das pressões junto ao fundo de dissipadores de energia por ressalto hidráulico. PhD Thesis. Federal University of Rio Grande do Sul; 2011.

[16] Souza PEA. Bacias de dissipação por ressalto hidráulico com baixo número de Froude - análise das pressões junto ao fundo da estrutura. MSc Thesis. Federal University of Rio Grande do Sul; 2012.

[17] Nóbrega JD. Metodologia teórica e experimental para determinação das características do ressalto hidráulico clássico. MSc Thesis. University of São Paulo; 2014.

[18] Peterka AJ. Hydraulic design of stilling basins and energy dissipators. Engineering Monograph No. 25, 8th ed. United States Department of the Interior, Bureau of Reclamation; 1984.

[19] Marques MG, Drapeau J, Verrette JL. Flutuação de pressão em um ressalto hidráulico. Revista Brasileira de Recursos Hídricos 1997; 2, 45-52.

[20] Hager WH. Energy Dissipators and Hydraulic jump. Water Science and Technology Library, Vol. 8. The Netherlands: Kluwer Academic Publishers; 1992.

[21] Bakhmeteff BA, Matzke AE. The hydraulic jump in terms of dynamic similarity. Transactions of the American Society of Civil Engineers 1936; 101 630-647.

[22] Simões ALA, Schulz HE, Porto RM. Simulação numérica e verificação experimental da posição da superfície livre de um ressalto hidráulico em um canal retangular. In: IAHR, XXIV Congresso Latinoamericano de Hidráulica, November 2010, Punta del Este, Uruguay.

[23] Simões ALA, Porto RM, Schulz HE. Superfície livre de escoamentos turbulentos em canais: vertedores em degraus e ressalto hidráulico. Revista Brasileira de Recursos Hídricos 2012; 17 125-139.

[24] Nóbrega JD, Schulz HE, Simões ALA, Porto RM. Measurement of turbulence parameters in hydraulic jumps using ultrasonic sensors and their correlation with macro-

scopic flow parameters. In: COBEM 2013: 22nd International Congress of Mechanical Engineering, 3-7 November 2013, Ribeirão Preto, Brazil.

[25] Nóbrega JD, Schulz HE, Zhu DZ. Free surface detection in hydraulic jumps through image analysis and ultrasonic sensor measurements. In: Chanson H, Toombes L. (eds.) Hydraulic Structures and Society - Engineering Challenges and Extremes: proceedings of the 5th IAHR International Symposium on Hydraulic Structures, ISHS2014, 25-27 June 2014, Brisbane, Australia.

[26] Simões ALA. Considerações sobre a hidráulica de vertedores em degraus: metodologias adimensionais para pré-dimensionamento. MSc thesis. University of São Paulo; 2008.

[27] Simões ALA, Schulz HE, Porto RM. Stepped and smooth spillways: resistance effects on stilling basin lengths. Journal of Hydraulic Research 2010; 48 329–337.

[28] Schulz HE, Simões ALA. Desenvolvimento da superfície livre em escoamentos aerados: analogia com leis básicas de transferência. Report I/II/11; 2011. Available from: http://stoa.usp.br/hidraulica/files/-1/16207/Schulz_H.E._Simões_A.L.A_LTR-Relatório+I_II_11.pdf (acessed 21 July 2014).

[29] Schulz HE, Simões ALA. Desenvolvimento da superfície livre em escoamentos aerados: analogia com leis básicas de transferência. Revista Brasileira de Recursos Hídricos 2013; 18 35–44.

[30] Schulz HE, Lobosco RJ, Simões ALA. Multiphase analysis of entrained air in skimming flows along stepped chutes. In: Hogge M, Van Keer R, Dick E, Malengier B, Slodicka M, Béchet E, Geuzaine C, Noels, L, Remacle J-F. (eds.) ACOMEN 2011: proceedings of the Fifth International Conference on Advanced Computational Methods in Engineering, 14-17 November 2011, Liège, Belgium. Université Catholique de Louvain/Université de Liège/Universiteit Gent: Leuven.

[31] Schulz HE, Lopes Júnior GB, Simões ALA, Lobosco RJ. One dimensional turbulent transfer using random square waves – scalar/velocity and velocity/velocity interactions. In: Schulz HE, Simões ALA, Lobosco RJ. (eds.) Hydrodynamics – Advanced Topics. 1 ed. InTech; 2011. p3-34.

[32] Schulz HE, Simões ALA, Janzen JG. Statistical approximations in gas-liquid mass transfer. In: Komori S, McGillis W, Kurose R. (eds.) GTWS 2010: proceedings of the 6th International Symposium on Gas Transfer at Water Surfaces 2010, 17-21 May 2010, Kyoto, Japan. Kyoto: Kyoto University Press; 2011.

[33] Schulz HE, Janzen JG. Concentration fields near air-water interfaces during interfacial mass transport: oxygen transport and random square wave analysis. Brazilian Journal of Chemical Engineering 2009; 26 527–536.

[34] Hughes WC, Flack JE. Hydraulic jump properties over a rough bed. Journal of Hydraulic Engineering 1984; 110 1755–1771.

[35] Bhutto HBG. Hydraulic jump control and energy dissipation. PhD thesis. Institute of Irrigation and Drainage Engineering, Mehran University of Engineering and Technology; 1987.

[36] Li C-F. Determining the location of hydraulic jump by model test and HEC-2 flow routing. MSc thesis. College of Engineering and Technology, Ohio University; 1995.

[37] Ead SA, Rajaratnam N. Hydraulic jumps on corrugated beds. Journal of Hydraulic Engineering 2002; 128 656–663.

[38] Evcimen TU. The effect of prismatic roughness elements on hydraulic jump. MSc thesis. The Graduate School of Natural and Applied Sciences, Middle East Technical University; 2005.

[39] Carollo FG, Ferro V, Pampalone V. Hydraulic jumps on rough beds. Journal of Hydraulic Engineering 2007; 133 989–999.

[40] Gandhi S, Yadav V. Characteristics of supercritical flow in rectangular channel. International Journal of Physical Sciences 2013; 8 1934–1943.

[41] Ayanlar K. Hydraulic jump on corrugated beds. MSc thesis. Middle East Technical University; 2004.

[42] Salehian S, Bajestan MS, Jahromi HM, Kashkooli H, Kashefipour SM. Hydraulic jump characteristics due to natural roughness. World Applied Sciences Journal 2011; 13 1005–1011.

[43] Souza PMBM. Estudo da dissipação de energia por ressalto hidráulico a jusante de descarregadores não convencionais. University of Porto; 2011.

[44] Izadjoo F, Bejestan MS. Corrugated bed hydraulic jump stilling basin. Journal of Applied Sciences 2007; 7 1164–1169.

[45] Rajaratnam, N. Hydraulic Jumps, in: Chow, VT. (ed) Advances in Hydroscience v.4, New York: Academic Press; 1967, 197-280.

[46] Lencastre A. Hidráulica Geral. Lisbon: Author Edition; 1996.

[47] Murzyn F, Chanson H. Free-surface fluctuations in hydraulic jumps: experimental observations. Experimental Thermal and Fluid Science 2009; 33 1055–1064.

Chapter 5

Lagrangian Hydrodynamics, Entropy and Dissipation

Massimo Materassi

Additional information is available at the end of the chapter

http://dx.doi.org/10.5772/59319

1. Introduction

In this Chapter we deal with a formalism that is both fundamental and an apparently "niche product": I myself confess to have taken the chance of Editors' kind invitation to go deeper into this subject. The subject I am speaking about is the *Lagrangian Formulation* (LF) of Fluid Dynamics, in particular its theoretical, brutally analytical application to the case of fluids with *dissipation*.

The reasons due to which treating Fluid Dynamics via LF is not "so popular" are that it involves equations of motion which are very complicated (apparently, "uselessly complicated"), and that squeezing practical results out of it seems to be desperately hard. Moreover, performing measurements from the "Lagrangian point of view" is not common, and requires to leave a probe be transported by the flow: it is often easier to realize a station instead, so to put everything in the other way we know to describe fluids, the "Eulerian point of view".

The Eulerian Formulation (EF), also indicated as local formulation because it treats the bulk properties of a fluid as properties of the space itself, appears to be much more popular than its "first brother" the LF. The EF is "more popular" because its equations of motion appear much simpler, and because their implementation on a calculator that moves its imagination on a 1D, 2D or 3D grid sounds much easier (still, the celebrated Navier-Stokes Equation, its key statement, is one of the most controversial in the community as far as "the knowledge of all its possible solutions" is concerned).

It is really complicated to find textbooks in which *non-ideal fluids* are described in the LF, and this is because if for dissipation free fluids the LF may sound cumbersome, when it comes to dissipation complications grow even harder.

The two formulations quoted have specific applications, and the point is that traditionally the applications at best represented in EF have been prevailing. As we are going to point out rigorously few lines below, the LF is following the material flow, while EF is best describing what is seen by an observer as she sits at a point $\vec{x}$ in space, and makes measurements letting the fluid matter passing by.

Different kinds of motion can be optimally studied in each different one of the two formalisms, depending on the actual nature of the motion at hand. In particular, material regions of fluid showing a highly correlated motion, namely *coherent structures* with all their particles "going together" (see Figure 1), are best described by following them as they move throughout the space: for this kind of modes the LF may have great value.

Figure 1. Clouds in the sky indicate a local steep gradient of refraction index of the air, due to the presence of water. As a cloud appears like a "body", i.e. a complex of matter moving together coherently, its motion, or the motion of the region of air containing it, may be best represented in the Lagrangian Formalism. Unfortunately, the non-linear essence of those coherent structures (as also vortices, or current sheets in plasmas) has always been a serious difficulty to study them analytically.

Instead, those perturbations that are travelling across the fluid without any real matter transfer, i.e. mechanical *waves*, are well studied as a classical field theory, in the EF (see Figure 2).

As the scientific literature witnesses, while an entire zoology of waves, both linear and non-linear, has been developed for fluids and plasmas, quite much less has been done for coherent structures (Chang 1999), and this is accompanied by the much larger diffusion EF studies than LF ones.

In the courses of dynamics of continua taught at the university, fluid systems are first described in the LF, or at least from the Lagrangian viewpoint (Rai Choudhuri, 1998); then, as soon as

http://dx.doi.org/10.5772/59319

Figure 2. Some mechanical waves propagating along the surface of a fluid. The description of such modes of the continuum is traditionally (and better) done in the Eulerian Formalism, where the physical quantities locally describing the motion of a fluid are conceived as local space properties, i.e. classical fields. In waves, indeed, matter does not propagate itself, and prevalently linear terms appear in the equations of motion. Picture by "hamad M", on Flickr, at the webpage https://www.flickr.com/photos/meshal/.

"serious equations" appear, the use of the EF prevails, because of the aforementioned difficulties. However, the fact that the language of the LF is chosen to initiate students to dynamics of continua is not an oddity, rather it reflects how this framework portrays continuous systems more understandably for people having been dealing with discrete systems up to that moment.

Continua described in the LF appear as very straightforward generalizations of systems formed by material points, and their behaviour sounds much more understandable at first glance keeping in mind that the discrete system dynamics looks like. This property of the LF has not a "didactic" value only, but rather can render highly transparent physical relationships between various scales of the theory, the interaction among which is the essence of dissipation. This is, in extremely few words, why I have chosen to treat this subject in the present Chapter.

In this Chapter a special care is devoted to examining how the results about dissipation, usually presented in the Eulerian framework, appear in the LF. Processes involved in dissipation are the ones keeping trace of the granular nature of matter, and a very natural way of seeing this is to describe the fluid in the Lagrangian Formulation. Actually, the postulates of the LF must be criticized right in vision of the granularity of matter, as reported in § 6. For the moment being, let's just consider the LF as non-in-conflict with matter granularity.

In § 2 the fundamental tools of the LF are proposed, together with the physical sense of the idea of fluid parcel, and the relationship of this with the particles of matters forming the continuum. In this § the fluid geometry, kinematics and mass conservation are sketched.

The application of the LF tools to fluids without dissipation is then given in § 3: in the absence of friction the fluid is treated as a Hamiltonian system, and here the dynamics appears in LF as inherited from that of point particles. The parcel variables of the LF are regarded as a centre-of-mass versus relative variables decomposition of the discrete analytical mechanics. As an aside of this interpretation, § 4 is used to clarify the differential algebraic properties of the entropy of the fluid encoding the degrees of freedom of the microscopic particles forming the parcel.

The granular nature of matter will enter in dynamics only in § 5, where dissipative fluids are treated in the LF: the dissipative terms completing the ideal equations of motion found in the Hamiltonian limit are written, and the Poisson algebra describing the non-dissipative fluid is completed with the introduction of a metric bracket, giving rise to the metriplectic algebra for dissipative fluid in LF.

Conclusions given in § 6 do concern the concept of parcel, and of postulates given in § 2, with respect to turbulence and matter granularity; then, applications and possible developments of the LF are traced.

2. Parcels versus particles

The LF describes a fluid by representing its motion as the evolution, throughout space, of the physical domain occupied by its matter (Bennett, 2006).

At a certain time t, let the locus of points in R^3 occupied by the fluid matter be indicated as $D(t)$; if the description of the motion starts at the time t_0, so that this is the initial time, let the initial material domain occupied by the fluid be indicated as D_0. The key question is: how to assign the fluid configuration at the time t ? Not only one has to assign the set $D(t) \subseteq R^3$ of the fluid domain: also, the distribution of matter within it must be given, and the instantaneous velocity of each material point of it. This is a big job, that is rendered possible by choosing to work at the scale at which the system appears as a continuum, and the "individual" motions of the particles composing it may be disregarded as important.

What is done in the LF is to imagine the fluid domain subdivided into "an infinite" number of "infinitesimal" portions, referred to as *parcels* that are still formed by a thermodynamic number of particles. The size of the parcel is chosen so that the *bulk macroscopic quantities* with which the fluid is described (velocity, mass density, tension…) are constant within it, but a sufficient number of elementary particles of the fluid is included, so that no extreme fluctuations (i.e. no sub-fluid effects as those described in Materassi et al. (2012)) can be appreciated. Parcels sized in this way are indicated as *macroscopic infinitesimal*, indeed, all the extensive quantities pertaining to the parcel (mass, energy, momentum) will be indicated with differentials.

At the initial time, this subdivision is snapshot stating that we have one parcel, of volume d^3a, at each point $\vec{a} \in D_0$. Those parcels are imagined to move, as the fluid evolves, with some constraints that are the foundations of the continuous approximation itself, and restrict very precisely the realm of application of the whole scheme. In particular, the following requirements are met:

1. *parcel identity conservation*: parcels move without mixing their matter with each other. They can be deformed, stretched, their volume does change during the evolution of the fluid, but they do not exchange matter, they do not fragment, they do not crumble into smaller entities;

2. the connectedness and smoothness of D_0 does not change as it evolves into $D(t)$, at no time: not only the set of continuous matter remains "all one" without being cut, also no cusps, spikes, "hairy" non-differentiable regions appear in it. In particular, in topological terms, one should state

$$\dim D(t) = \dim D_0, \tag{1}$$

being dim the Hausdorff dimension of fluid matter.

Of course, it is necessary to remind that this LF can be improved, on the side of treatment of irregular motions, by adding noise terms to its otherwise deterministic and C^∞ equations (Materassi & Consolini, 2008). Nevertheless, here we are only interested in giving the general directions of the deterministic LF dynamics.

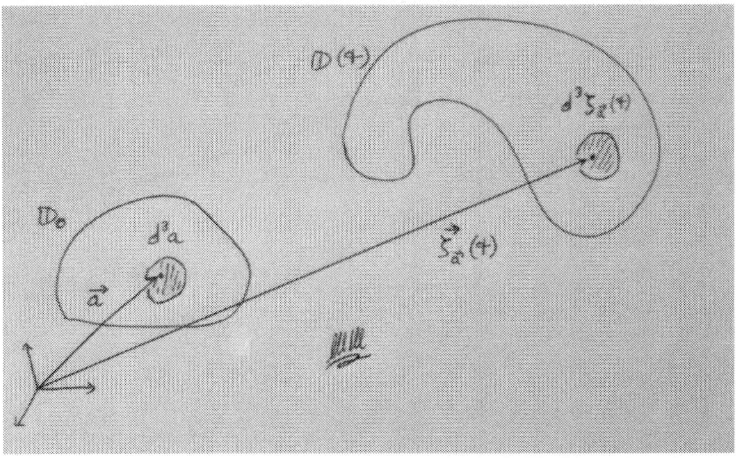

Figure 3. The transformation from the initial configuration of the fluid D_0 into the one assumed at the generic time $D(t)$. The parcel, initially located at the position $\vec{a}$ assumes position $\zeta_{\vec{a}}(t)$, while its initial infinitesimal volume d^3a is transformed into the evolved $d^3\zeta_{\vec{a}}(t)$ (note the use of $\zeta(\vec{a}, t)$ instead of $\zeta_{\vec{a}}(t)$ in the text).

The two requirements stated before are translated into a conceptually very simple statement, in practice giving rise to complicated equations: *the evolution map $D_0 \mapsto D(t)$ is represented by a diffeomorphism at every timet.* In this way, each neighbourhood of every point $\vec{a} \in D_0$ is mapped in an invertible way into a unique neighbourhood of the point$\vec{\zeta}(\vec{a}, t) \in D(t)$, and the identity of parcels is conserved. Not only, being the diffeomorphism a C^{∞} map, the motion of the fluid is also smooth, as required.

The geometry and kinematic of the fluid are all contained in the diffeomorphism $\vec{a} \mapsto \vec{\zeta}(\vec{a}, t)$. It is natural to use $\vec{\zeta}(\vec{a}, t)$ itself as the dynamical variable of the theory, so that the LF is essentially a the field theory of $\vec{\zeta}(\vec{a}, t)$ on the 3D domain D_0 : the vector $\vec{\zeta}(\vec{a}, t)$ is the position of the $\vec{a}$-th parcel in the physical space R^3, while $\vec{a}$ is the label of the parcel, but it represents its initial position too:

$$\vec{\zeta}\left(\vec{a}, t_0\right) = \vec{a} \tag{2}$$

(see Figure 3). Even if (2) underlines how the vectors $\vec{\zeta}$ and the vectors $\vec{a}$ do "live" in the same physical space, it is useful to indicate their components with different labels: we will use Greek indices for$\vec{\zeta}$, writing its components as ζ^{α}, while Latin indices will label the components of $\vec{a}$, as a^i. Summation convention over repeated indices in contravariant position will be understood, so that scalar products will be written as:

$$\vec{\zeta} \cdot \vec{\xi} = \zeta^{\alpha} \xi_{\alpha}, \quad \vec{a} \cdot \vec{b} = a^i b_i. \tag{3}$$

The velocity of the $\vec{a}$-th parcel will simply be $\partial_t \vec{\zeta}(\vec{a}, t)$, even if this notation might be misleading: this is *a total derivative with respect to time,* because the "index" $\vec{a}$in $\vec{\zeta}(\vec{a}, t)$ does not move at all, but is just "the name" of the parcel.

Another geometrical-kinematical object to be defined in LF is the *Jacobian matrix of the diffeomorphism $\vec{a} \mapsto \vec{\zeta}(\vec{a}, t)$,* indicated as$J(\vec{a}, t) = \partial \vec{\zeta}(\vec{a}, t) / \partial \vec{a}$. Its components, and the components of its inverse matrix, read

$$J_i^{\alpha} = \frac{\partial \zeta^{\alpha}}{\partial a^i}, \quad \left(J^{-1}\right)_{\alpha}^{i} = \frac{\partial a^i}{\partial \zeta^{\alpha}}. \tag{4}$$

The determinant of the Jacobian is indicated as:

$$J = \det J. \tag{5}$$

Clearly, $J(\vec{a}, t_0) = 1 \quad \forall \quad \vec{a} \in D_0$. Last, but not least, some *time-semigroup property* of the diffeomorphism must hold: $\vec{\zeta}(\vec{a}, t_1 + t_2) = \vec{\zeta}(\vec{\zeta}(\vec{a}, t_1), t_2)$.

http://dx.doi.org/10.5772/59319

The use of $J(\vec{a}, t)$ renders it straightforward, in practice built-in, the *mass conservation* in LF: due to the parcel identity conservation *the parcel mass will be conserved itself*, so that, if $\rho_0(\vec{a})$ is the initial mass density of the $\vec{a}$-th parcel, one may write the initial mass of the parcel as $dm(\vec{a}, t_0) = \rho_0(\vec{a}) d^3 a$. On the other hand, at time t the same $\vec{a}$-th infinitesimal mass will be written as $dm(\vec{a}, t) = \rho(\vec{a}, t) d^3 \zeta(\vec{a}, t)$, being $\rho(\vec{a}, t)$ the $\vec{a}$-th mass density at a generic time, and $d^3 \zeta(\vec{a}, t)$ the infinitesimal volume of the parcel at the same time. The parcel mass conservation hence reads

$$\rho_0(\vec{a}) d^3 a = \rho(\vec{a},t) d^3 \zeta(\vec{a},t) , \tag{6}$$

and since the relationship between $d^3 a$ and $d^3 \zeta(\vec{a}, t)$ is given by $d^3 \zeta(\vec{a}, t) = J(\vec{a}, t) d^3 a$, so that one has:

$$\rho(\vec{a},t) = \frac{\rho_0(\vec{a})}{J(\vec{a},t)}. \tag{7}$$

The relationships (6) and (7) are all that is needed, in the LF, to let the mass conservation be adequately represented (those relationships are turned into the continuity equation of the EF).

Once the position of the parcel in R^3 and its mass density have been described, the reality of the parcel as composed by a thermodynamic number of *microscopic particles* must be considered. Indeed, even if the parcel identity conservation and the smoothness of the fluid evolution render $\vec{\zeta}(\vec{a}, t)$ sufficient to describe the configuration of the fluid at any time, while $\rho(\vec{a}, t)$, related to the initial mass density $\rho_0(\vec{a})$ as equation (7) prescribes, completes also the mass geometry and mass conservation description, there are *dynamical effects*, that will be examined in §§ 3, 4 and 5, through which the granular nature of matter do come into the play.

Let's hence suppose that the $\vec{a}$-th parcel is formed by $N(\vec{a})$ *particles*, each of them located in the space by a position vector $\vec{r}_{I=1,\dots,N(\vec{a})}$ that belongs to the neighbourhood centred in $\vec{\zeta}(\vec{a}, t)$ and of measure $J(\vec{a}, t) d^3 a$. When these particles are considered, it is no surprise to state that the vector $\vec{\zeta}(\vec{a}, t)$, understood as "the parcel position", may be defined as the position of the centre-of-mass of the $N(\vec{a})$ particles in the parcel (Arnold, 1989), see Figure 4:

$$\vec{\zeta}(\vec{a},t) = \frac{1}{N(\vec{a})} \sum_{I=1}^{N(\vec{a})} \vec{r}_I. \tag{8}$$

In the foregoing formula one supposes that all the particles have the same mass, so that a mere arithmetical average of their positions can be taken to define the centre-of-mass. In order to complete the set of centre-of-mass variables, let's consider there also will be a parcel momentum density

$$\vec{\pi}(\vec{a},t) = \rho_0 \partial_t \vec{\zeta}(\vec{a},t) = \frac{1}{d^3 a} \sum_{I=1}^{N(\vec{a})} \vec{p}_I, \tag{9}$$

that is defined more rigorously in § 3, equation (18) (in (9) $\vec{p}_I$is the linear momentum of the I-th particle in the $\vec{a}$-th parcel).

By considering the definitions (8) and (9), out of the $6N(\vec{a})$ dynamical variables $(\vec{r}_I, \vec{p}_I)$ that analytical mechanics would attribute to the $N(\vec{a})$ particles, $6(N(\vec{a})-1)$remain, those that would be referred to as *relative-to-the-centre-of-mass*, or simply *relative variables*. Due to how huge $N(\vec{a})$ is, and considering $N(\vec{a})-1 \cong N(\vec{a})$, instead of defining relative independent variables in subtle complicated ways (Lusanna & Materassi, 2000), one decides to juxtapose to the variables $(\vec{\zeta}, \vec{\pi})$ the *Equilibrium Statistical Mechanics* of $N(\vec{a})$ particles of the species forming the fluid. In a sense, this is as treating the relative variables as *microscopic degrees of freedom treated statistically* (Materassi et al., 2012). As it happens for ideal gases, those namely $6(N(\vec{a})-1)$ dynamical variables may be effectively described via *two state functions* that should be the poor man's version of the Statistical Mechanics of the ponderous $6N(\vec{a})$-dimensional phase space. In the undergraduate student's thermodynamics, a gas can be described, for instance, via its *density* ρ and its *entropy* S, and so will be for the pointlike masses forming our parcel: the parcel density has already been assigned through (7), what remains is to equip the field configuration of the fluid in the LF by some entropic field, i.e. the *mass-specific entropy density* $s(\vec{a}, t)$, defined such that the parcel's infinitesimal entropy reads:

$$dS(\vec{a},t) = \rho(\vec{a},t)s(\vec{a},t)d^3\zeta(\vec{a},t) = \rho_0(\vec{a})s(\vec{a},t)d^3a. \tag{10}$$

The set of microscopic particles forming the $\vec{a}$-th parcel is hence a macroscopically infinitesimal cloud of position $\vec{\zeta}(\vec{a}, t)$ and total momentum $\vec{\pi}(\vec{a}, t)d^3a$.

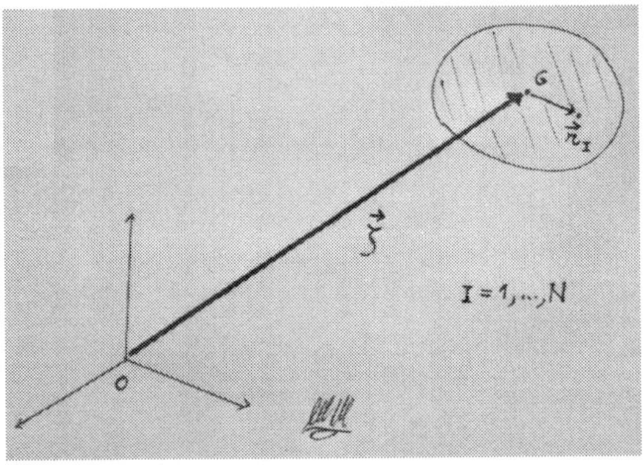

Figure 4. The parcel of position $\vec{\zeta}$ and the N particles forming it, with position relative-to-the-parcel-centre-of-mass indicated with $\vec{r}_I$. The centre-of-mass of the parcel is the point G, of position $\vec{\zeta}$ in the space.

The "statistical buzz" of those $N(\vec{a})$ pointlike masses is encoded in the thermodynamics of $J(\vec{a},\ t)$ and $s(\vec{a},\ t)$. Since $J(\vec{a},\ t)$ is already encoded in $\vec{\zeta}(\vec{a},\ t)$, one concludes that the functionally independent fields $\vec{\zeta}(\vec{a},\ t)$, $\vec{\pi}(\vec{a},\ t)$and $s(\vec{a},\ t)$ describe completely the field configuration, and they will be referred to as *parcel variables*.

3. Ideal fluids in Lagrangian formalism

All in all, we have established that the fluid geometry, kinematics and mass geometry may be represented in parcel variables through the parcel variables $\vec{\zeta}(\vec{a},\ t)$, $\vec{\pi}(\vec{a},\ t)$and $s(\vec{a},\ t)$. Now it is time to describe the interactions to which those quantities undergo, determining the dynamics of the fluid as represented in parcel variables.

The dynamics of the fields $\vec{\zeta}(\vec{a},\ t)$ and $\vec{\pi}(\vec{a},\ t)$ describe the "collective" evolution of the matter forming the $\vec{a}$-th parcel: this will be determined by the forces external to the parcel (Feynman, 1963) acting on it. One may consider forces due to potentials that are not due to the fluid, for instance gravitation; there will be also the forces exerted on the parcel by the nearby parcels. The latter are subdivided into the "conservative" force due to the elasticity of the continuum, that will be encoded in *pressure*, and the "dissipative" force that parcels exert on the nearby ones by rubbing each other, i.e. *friction*. Both are expected to depend on gradients with respect to the label, because these are all due to the nearby parcels.

While external potential forces are exquisitely pertaining the collective position $\vec{\zeta}(\vec{a},\ t)$, forces exchanged with the nearby parcels are encoded in the thermodynamics of the particles forming the parcel: they may be expressed through the internal energy density $\rho_0 U\left(\frac{\rho_0}{J},\ s\right)$ of the fluid.

In this § the dissipation free limit will be studied, excluding for the moment the friction between nearby parcels, treated in § 5.

As elegantly suggested by Goldstein (2002), when dissipation is ruled out a mechanical system is expected to undergo an *Action Principle*, and this will be the case for the field theory of parcel variables. In order to write down the mechanical action of the fluid in LF, a resume is done of the forms of energy attributed to the parcel.

The parcel has a collective motion throughout the space, conferring it an amount of kinetic energy

$$dE_{\text{kin}} = \frac{\rho_0 \dot{\zeta}^2}{2} d^3 a, \tag{11}$$

where $\dot{\zeta}^2 = |\partial_t \vec{\zeta}|^2$ is intended. Then, assuming there exist potentials external to the fluid, these are written as:

$$dV = \rho_0 \phi(\vec{\zeta}) d^3 a. \tag{12}$$

The parcel possesses also the internal thermodynamic energy of the particles forming it, that we have already mentioned before and that here we indicate as:

$$dE_{\text{therm}} = \rho_0 U\left(\frac{\rho_0}{J}, s\right) d^3 a. \tag{13}$$

When equations (11), (12) and (13) are put all together, one may attribute *an infinitesimal Lagrangian* to the $\vec{a}$-th parcel, that reads:

$$dL = \rho_0 \left[\frac{\dot{\zeta}^2}{2} - \phi(\vec{\zeta}) - U\left(\frac{\rho_0}{J}, s\right)\right] d^3 a, \tag{14}$$

so that the Lagrangian of the whole system is simply the dL in (14) integrated on the initial volume D_0:

$$L[\vec{\zeta}, s] = \int_{D_0} \rho_0 \left[\frac{\dot{\zeta}^2}{2} - \phi(\vec{\zeta}) - U\left(\frac{\rho_0}{J}, s\right)\right] d^3 a. \tag{15}$$

The mechanical action for the fluid in LF is the aforementioned functional integrated in time along the interval $[t_0, t_f]$ along which we are interested in studying the system:

$$A[\vec{\zeta}, s] = \int_{t_0}^{t_f} dt' \int_{D_0} \rho_0 \left[\frac{\dot{\zeta}^2}{2} - \phi(\vec{\zeta}) - U\left(\frac{\rho_0}{J}, s\right)\right] d^3 a. \tag{16}$$

In Padhye (1998), those expressions are used to find the Euler-Lagrange equations for the natural motions of the system, that read:

$$\ddot{\zeta}_\alpha = -\frac{\partial \phi}{\partial \zeta^\alpha} + \frac{A^i_\alpha}{\rho_0} \frac{\partial}{\partial a^i}\left(\rho_0 \frac{\partial U}{\partial J}\right), \quad A^i_\alpha = \frac{1}{2} \varepsilon_{\alpha\kappa\lambda} \varepsilon^{imn} \frac{\partial \zeta^\kappa}{\partial a^m} \frac{\partial \zeta^\lambda}{\partial a^n}, \\ \dot{s} = 0. \tag{17}$$

http://dx.doi.org/10.5772/59319

In (17) the term $A_\alpha^i \frac{\partial}{\partial a^i}\left(\rho_0 \frac{\partial U}{\partial J}\right)$ is exactly the infinitesimal force insisting on the parcel due to the *adiabatic pressure* exerted on it by its nearby fellows: in particular, one should recognize the pressure undergone by the parcel as $p = -\rho_0 \frac{\partial U}{\partial J}$.

The second differential equation in (17) deserves some more explanation.

Varying the action A with respect to s, the effect of the absence of any derivative of it in dL appears in the final statement according to which it must be constant. This corresponds to the result that, in the dissipation free limit in which the Action Principle is applicable, the mass-specific entropy density of the $\vec{a}$-th parcel does not change. Actually, in this case the whole entropy of the parcel remains constant, as it can be seen from $dS(\vec{a}, t) = \rho_0(\vec{a})s(\vec{a}, t)d^3a$. The quantity $s(\vec{a}, t)$ turns out to be *a passively advected scalar*, driven throughout the space by the parcel that carries it.

The constancy of $s(\vec{a}, t)$ means adiabaticity (isoentropicity) of the parcel motion for the ideal fluid because the density at hand is supposed to play the role of mass-specific entropy density in the definition of the amount of internal energy dE_{therm} in (13).

Once the Lagrangian dynamics is defined through (14) and (15) it is possible to turn it into a Hamiltonian canonical framework through the Legendre transformation of $L[\vec{\zeta}, s]$ into the Hamiltonian $H[\vec{\zeta}, \vec{\pi}, s]$ (Goldstein, 2002).

First of all, a rigorous definition of the kinetic momentum $\vec{\pi}(\vec{a}, t)$ must be given as:

$$\pi_\beta(\vec{a}) = \frac{\delta L}{\delta \dot{\zeta}^\beta(\vec{a})}. \tag{18}$$

By calculating the functional derivative indicated in (18), the result (9) is obtained. Then, the calculation of the Hamiltonian functional of the ideal fluid in LF is straightforward:

$$H\left[\vec{\zeta}, \vec{\pi}, s\right] = \int_{D_0} d^3a \left[\frac{\pi^2}{2\rho_0} + \rho_0\phi\left(\vec{\zeta}\right) + \rho_0 U\left(\frac{\rho_0}{J}, s\right)\right]. \tag{19}$$

The Hamilton Equations of motion of the system have a canonical form $\dot{\zeta}^\alpha = \delta H / \delta\pi_\alpha$ and $\dot{\pi}_\beta = -\delta H / \delta\zeta^\beta$, that is:

$$\dot{\zeta}^\alpha = \frac{\pi^\alpha}{\rho_0}, \quad \dot{\pi}_\alpha = -\rho_0 \frac{\partial\phi}{\partial\zeta^\alpha} + A_\alpha^i \frac{\partial}{\partial a^i}\left(\rho_0 \frac{\partial U}{\partial J}\right),$$
$$\dot{s} = 0, \tag{20}$$

with the same definition for the symbol A_α^i as given in (17).

Equations (20) may be put in the form of a *Poisson algebra*, by defining the Poisson bracket

$$\{f,g\}_{\mathrm{LF}} = \int_{D_0} d^3a \left[\frac{\delta f}{\delta\zeta^\alpha(\vec{a})} \frac{\delta g}{\delta\pi_\alpha(\vec{a})} - \frac{\delta g}{\delta\zeta^\alpha(\vec{a})} \frac{\delta f}{\delta\pi_\alpha(\vec{a})} \right]; \tag{21}$$

then, the dynamics of any physical variable of the system is assigned as

$$\dot{f} = \{f,H\}_{\mathrm{LF}}. \tag{22}$$

In particular, if the Poisson brackets $\{\vec{\zeta}, H\}_{\mathrm{LF}}$, $\{\vec{\pi}, H\}_{\mathrm{LF}}$ and $\{s, H\}_{\mathrm{LF}}$ are calculated, it is easy to obtain the expressions (20).

4. The Casimir $S[s]$

An important observation should be done on the definition of $\{.,.\}_{\mathrm{LF}}$ in (21): indeed, despite the fact that the full field configuration of the system is given by the collection of three fields $(\vec{\zeta}, \vec{\pi}, s)$, the Poisson bracket only involves derivative with respect to $\vec{\zeta}$ and $\vec{\pi}$ only, so that for sure one has:

$$\{C,F\}_{\mathrm{LF}} = 0 \quad \forall \quad C[s], \ F[\vec{\zeta},\vec{\pi},s], \tag{23}$$

i.e. *any quantity $C[s]$ depending only on the mass-specific entropy density is in involution with any quantity at all.* Quantities which are in involution "with anything" are referred to as *Casimir invariants*: in the Poisson algebra $(O_{\mathrm{fluid}}^{\mathrm{LF}}, \{.,.\}_{\mathrm{LF}})$ of the fluid dynamics observables $O_{\mathrm{fluid}}^{\mathrm{LF}}$ in the LF composed by the Poisson bracket $\{.,.\}_{\mathrm{LF}}$ in (21), all the functionals of $s(\vec{a}, t)$ are Casimir invariants. In particular, the *total entropy of the fluid*

$$S[s] = \int_{D_0} \rho_0(\vec{a}) s(\vec{a},t) d^3a \tag{24}$$

is the most important one of them, as it will be stressed in § 5.

When an algebra $(O, \{.,.\})$ admits Casimir invariants it means that the Poisson bracket is *singular*, i.e. it is a bilinear application on observables giving a null result also for some non zero argument. Typically, Poisson brackets of this kind are not in the form (21): indeed, apparently that is the same expression $\{f(q,\,p),\,g(q,\,p)\} = \frac{\partial f}{\partial q}\cdot\frac{\partial g}{\partial p} - \frac{\partial g}{\partial q}\cdot\frac{\partial f}{\partial p}$ used in the canonical Hamiltonian systems with discrete variables. From Mathematical Physics we know that canonical brackets cannot have Casimirs, so there's something "missing" in here. The point is that, as far as the set $O^{\mathrm{LF}}_{\mathrm{fluid}}$ is the one of (smooth) functionals depending on the whole field configuration $(\vec{\zeta},\,\vec{\pi},\,s)$, *indeed* $\{.,\,.\}_{LF}$ *is not, strictly speaking, a canonical bracket*, due to the lack of $\frac{\delta}{\delta s(\vec{a})}$ derivatives. What can be said is that the Poisson bracket $\{.,\,.\}_{\mathrm{LF}}$ is canonical if restricted to the subset of $O^{\mathrm{LF}}_{\mathrm{fluid}}$ of functionals depending only on the centre-of-mass variables $\vec{\zeta}$ and $\vec{\pi}$, while this properties is lost if the microscopic degrees of freedom "within the parcel" are considered too.

When Casimir invariants appear it is generally because a Poisson algebra $(M,\,\{.,\,.\})$ undergoes a *process of reduction*.

Assume to have in the set M a Hamiltonian such that $\dot{f} = \{f,\,H\}$ for any element $f \in M$, and then observe H to be invariant under a certain group G of transformations, realized through its Lie algebra $\boldsymbol{g}$ on $(M,\,\{.,\,.\})$, so that $\{g,\,H\} = 0$ for any $g \in \boldsymbol{g}$. One defines the G-reduced algebra $(\boldsymbol{M}_{\mathrm{red}},\,\{.,\,.\}_{\mathrm{red}})$ as the set

$$\boldsymbol{M}_{\mathrm{red}} : \left\{ f \in \boldsymbol{M} \;\; / \;\; \left\{ g, f \right\} = 0 \;\; \forall \;\; g \in \boldsymbol{g} \right\} . \tag{25}$$

This sub-algebra $(\boldsymbol{M}_{\mathrm{red}},\,\{.,\,.\}_{\mathrm{red}})$ of all the G-invariant elements of $\boldsymbol{M}$ is still a Poisson bracket algebra, where $\{.,\,.\}_{\mathrm{red}}$ is the restriction of $\{.,\,.\}$ to $\boldsymbol{M}_{\mathrm{red}}$. By the definition of $\boldsymbol{M}_{\mathrm{red}}$ itself, it's clear that any $g \in \boldsymbol{g}$ realized on $\boldsymbol{M}$ is a Casimir for $\{.,\,.\}_{\mathrm{red}}$:

$$\left\{ f, g \right\}_{\mathrm{red}} = 0 \;\; \forall \;\; f \in \boldsymbol{M}_{\mathrm{red}}, \;\; g \in \boldsymbol{g} . \tag{26}$$

It may happen that, while $\{.,\,.\}$ is a canonical Poisson bracket, $\{.,\,.\}_{\mathrm{red}}$ is non-canonical (still, the system $\dot{f} = \{f,\,H\}_{\mathrm{red}}$ is Hamiltonian).

The map from the Lagrangian quantities in $O^{\mathrm{LF}}_{\mathrm{fluid}}$ to their correspondent Eulerian fields may be regarded as a reduction from $(O^{\mathrm{LF}}_{\mathrm{fluid}},\,\{.,\,.\}_{\mathrm{LF}})$ to the non-canonical Poisson algebra $(O^{\mathrm{EF}}_{\mathrm{fluid}},\,\{.,\,.\}_{\mathrm{EF}})$ (Morrison & Greene, 1980). $S[s]$ is then regarded as a Casimir invariant due to the reduction $(O^{\mathrm{LF}}_{\mathrm{fluid}},\,\{.,\,.\}_{\mathrm{LF}}) \mapsto (O^{\mathrm{EF}}_{\mathrm{fluid}},\,\{.,\,.\}_{\mathrm{EF}})$, where the transformations under which H is invariant is that of *relabeling transformations* (the Hamiltonian in (19) remains the same if the parcels are smoothly relabelled (Padhye & Morrison, 1996 a,b)). The original claim here is that this is not really the case: rather $S[s]$ is *already a Casimir* of $\{.,\,.\}_{\mathrm{LF}}$, as pointed before, and of course *it remains a Casimir* for $\{.,\,.\}_{\mathrm{EF}}$ too (Materassi, 2014). The author's opinion is that yes, the entropy comes out as a Casimir from a reduction process, of which, however, the algebra M_{red} is already $O^{\mathrm{LF}}_{\mathrm{fluid}}$: in fact, $\{.,\,.\}_{\mathrm{LF}}$ is *already non-canonical*.

The "original algebra" $(\mathcal{O}_{\text{fluid}}, \{.,.\})$that should be reduced to get $\left(\mathcal{O}^{\text{LF}}_{\text{fluid}}, \{.,.\}_{\text{LF}}\right)$ must be an algebra where the canonical couple $(\vec{\zeta}(\vec{a}), \vec{\pi}(\vec{a}))$ is completed with $6(N(\vec{a})-1)$ canonical variables describing the motion of the $N(\vec{a})$ particles of the $\vec{a}$-th parcel relative-to-the-centre-of-mass $\vec{\zeta}(\vec{a})$. The act-of-motion of the $\vec{a}$-th parcel in R^3, described by $(\vec{\zeta}(\vec{a}), \vec{\pi}(\vec{a}))$, is *invariant under some class of transformation of the microscopic state of the parcel's particles*, and this should drive the reduction $(\mathcal{O}_{\text{fluid}}, \{.,.\}) \mapsto \left(\mathcal{O}^{\text{LF}}_{\text{fluid}}, \{.,.\}_{\text{LF}}\right)$.

Even if the picture is rather clear, from a logical point of view, making $s(\vec{a})$ arise from microscopic reshuffling transformations appear out of reach yet.

5. Dissipative fluids in Lagrangian formalism

In the absence of dissipation, the "decoupling" of the mass-specific entropy density $s(\vec{a}, t)$ from the other parcel variable in Equations (17) means the decoupling of the microscopic degrees of freedom of the particles forming the parcel, encoded only in $s(\vec{a}, t)$. That could be expected, since *dissipation should be intended as the dynamical shortcut between the macroscopic and microscopic level of the theory in terms of force and energy exchange*: so, excluding these exchanges means decoupling of $s(\vec{a}, t)$ from the centre-of-mass variable $\vec{\zeta}(\vec{a}, t)$. One may also state that the *one-way direction* of this exchange is an aspect of the *Second Law of Thermodynamics* (Dewar et al., 2014).

When dissipation is not present, the microscopic degrees of freedom animating the particles *within the parcel* are as frozen, and their world participates to the motion of the nearby parcels only through the compression-dilatation effect recorded by the term $A^i_\alpha \frac{\partial}{\partial a^i}\left(\rho_0 \frac{\partial U}{\partial J}\right)$. When dissipation is at work, then two important facts take place:

1. the energy transfer from the macroscopic degrees of freedom $(\vec{\zeta}, \vec{\pi})$ to the microscopic ones encoded in s will render $\dot{s} \neq 0$ and, according to the Second Principle, make it grow;
2. the same interaction will impede the "conservative motion" of the parcel centre-of-mass $(\vec{\zeta}, \vec{\pi})$ via "friction terms" rising due to the particle action.

All in all, both the centre-of-mass motion of the parcel and its entropy will be influenced by the presence of dissipation.

The equations of motion of fluids in the LF in the presence of dissipative terms are not such a widespread subject to be treated, let me address the reader to the book by Bennett "Lagrangian Fluid Dynamics", where the equation of motion of a viscous incompressible fluid is obtained in parcel variables (Bennett, 2006).

Looking at the equations (20), one way of finding out how to modify them in order to introduce dissipation is to examine the corresponding equations in EF (the dissipative Navier-Stokes equations, see for instance Rai Choudhuri (1998)) and then work on how the various quantities change when they undergo the inverse map $\vec{\zeta} \mapsto \vec{a}$. This is the way in which the LF equations of motion with dissipation were obtained in Materassi (2014).

http://dx.doi.org/10.5772/59319

The relationship between $\vec{\zeta}$ and $\vec{\pi}$ remain the same as in (20) when dissipation is introduced, that is $\zeta^{\alpha} = \frac{\pi^{\alpha}}{\rho_0}$.

The equation of motion for the momentum $\vec{\pi}$ changes because *now friction forces appear*: these pertain to the relative motion of the parcel with respect to its neighbours, so that one expects (as in the viscous force between pointlike particles and a medium) these will depend on the velocity difference between neighbouring parcels. Not only: due to the smoothness of the map $\vec{a} \mapsto \vec{\zeta}$, it is clear that neighbouring parcels do correspond to neighbouring labels, and so the friction forces will depend on the gradients of velocities, i.e. on $\vec{a}$-gradients of the field $\vec{\pi}(\vec{a}, t)$.

It is possible to show that, provided the stress tensor of the fluid appears in the EF as reported in Materassi et al. (2012), it is possible to write the equation of motion for π_{α} in the presence of dissipation as:

$$\dot{\pi}_{\alpha} = -\rho_0 \frac{\partial \phi}{\partial \zeta^{\alpha}} + A^{i}_{\alpha} \frac{\partial}{\partial a^{i}} \left(\rho_0 \frac{\partial U}{\partial J} \right) + J \Lambda_{\alpha\beta\gamma\delta} \nabla^{\beta} \nabla^{\gamma} \left(\frac{\pi^{\delta}}{\rho_0} \right). \tag{27}$$

In (27) the tensor $\Lambda_{\alpha\beta\gamma\delta}$ is constant, and reads:

$$\Lambda_{\alpha\beta\gamma\delta} = \eta \left(\delta_{\alpha\delta}\delta_{\beta\gamma} + \delta_{\beta\delta}\delta_{\alpha\gamma} - \frac{2}{3}\delta_{\gamma\delta}\delta_{\alpha\beta} \right) + \nu \delta_{\gamma\delta}\delta_{\alpha\beta}. \tag{28}$$

The coefficients η and ν are the two kinds of viscosity (compressible and incompressible, respectively) already quoted in Materassi & Tassi (2012 a) and in Materassi et al. (2012).

The derivative indicated as ∇^{β} is a differential operator acting on fields that depend on the label $\vec{a}$, and is constructed *through the chain rule* as:

$$\nabla^{\beta} = \frac{\partial a^{i}}{\partial \zeta_{\beta}} \frac{\partial}{\partial a^{i}} = \left(J^{-1} \right)^{i}_{\beta} \frac{\partial}{\partial a^{i}}, \tag{29}$$

being the matrix J^{-1} the Jacobian of the inverse map $\vec{\zeta} \mapsto \vec{a}$.

The last equation in (20) must be modified in order to include the dissipative terms in the entropy evolution.

In particular, in order to understand which terms should be included, it is of some use to consider that the entropy variation always corresponds, within quasi-equilibrium processes (Zemansky & Dittman, 1996), to *the ratio between the heat transferred and the temperature at which this transfer takes place*. Then, one has just to think about which are the processes transferring heat to the parcel.

Two sources of heat must be considered: there is a contribution given by the work dissipated through viscosity along the surface of the parcel, that should correspond to the force terms $J\Lambda_{\alpha\beta\gamma\delta}\nabla^{\beta}\nabla^{\gamma}\left(\frac{\pi^{\delta}}{\rho_0}\right)$; then, there is the heat flowing into the parcel, still across its surface, due to the finiteness of the thermal conductivity κ of the fluid.

The temperature at which these heat transfers take place is the temperature of the $\vec{a}$-th parcel at time t, that may as well be obtained by the internal energy U thanks to the relationship $T=\frac{\partial U}{\partial s}$.

In Bennett (2006) and in Materassi (2014) both the heat transfer terms have been obtained from the corresponding terms in the EF equations (as these are reported, e.g., in Morrison (1984) and Materassi & Tassi (2012 a)) through the inverse map $\vec{\zeta}\mapsto\vec{a}$:

$$\delta Q_{work} = \frac{\Lambda_{\alpha\beta\delta\gamma}\partial^{\alpha}v^{\beta}\partial^{\gamma}v^{\delta}}{\rho T}dt, \quad \delta Q_{conduction} = \frac{\kappa}{\rho T}\partial^2 T dt. \tag{30}$$

In (30) all the quantities are local fields, i.e. Eulerian variables in $O_{\text{fluid}}^{\text{EF}}$, while $\vec{\partial}$ is the gradient with respect to the space position, of components ∂^{α}, and $\vec{v}$ is the Eulerian field of velocities, of components v^{β}. The convention $\partial^{\alpha}\partial_{\alpha}=\partial^2$ was used.

After some relatively easy, but lengthy, algebra on (30), one ends up with the following equation of motion of the mass-specific entropy density:

$$\dot{s} = \frac{J}{\rho_0 T}\Lambda_{\alpha\beta\delta\gamma}\nabla^{\alpha}\left(\frac{\pi^{\beta}}{\rho_0}\right)\nabla^{\gamma}\left(\frac{\pi^{\delta}}{\rho_0}\right) + \frac{\kappa J}{\rho_0 T}\nabla^{\eta}\nabla_{\eta}T. \tag{31}$$

We are now in the position to provide the reader with the system of equations generalizing (20) in the case of viscous fluids:

$$\begin{gathered}\dot{\zeta}^{\alpha} = \frac{\pi^{\alpha}}{\rho_0}, \quad \dot{\pi}_{\alpha} = -\rho_0\frac{\partial\phi}{\partial\zeta^{\alpha}} + A_{\alpha}^{i}\frac{\partial}{\partial a^{i}}\left(\rho_0\frac{\partial U}{\partial J}\right) + J\Lambda_{\alpha\beta\gamma\delta}\nabla^{\beta}\nabla^{\gamma}\left(\frac{\pi^{\delta}}{\rho_0}\right), \\ \dot{s} = \frac{J}{\rho_0 T}\Lambda_{\alpha\beta\delta\gamma}\nabla^{\alpha}\left(\frac{\pi^{\beta}}{\rho_0}\right)\nabla^{\gamma}\left(\frac{\pi^{\delta}}{\rho_0}\right) + \frac{\kappa J}{\rho_0 T}\nabla^{\eta}\nabla_{\eta}T.\end{gathered} \tag{32}$$

In general, equations as (32) are extremely complicated to treat because of the very non-linear way in which the unknown functions, i.e. the fields $\vec{\zeta}(\vec{a}, t)$, $\vec{\pi}(\vec{a}, t)$ and $s(\vec{a}, t)$, participate in the various operators. For instance, one should underline how the gradients of $\vec{\zeta}(\vec{a}, t)$ appear in A_{α}^{i} through the inverse Jacobians (see (I. 17)), or in the determinants J, not to mention the operators ∇^{η}, especially when they are squared, so that, for instance, one has:

http://dx.doi.org/10.5772/59319

$\nabla^{\beta}\nabla_{\beta}T=\left(J^{-1}\right)^{i}_{\beta}\frac{\partial}{\partial a^{i}}\left[\left(J^{-1}\right)^{j\beta}\frac{\partial}{\partial a^{j}}T\right]$ in the entropy evolution equation. The intriguing thing is that this high complexity of the LF equations is due to the necessity of taking into account of how the fluid parcel deforms during its motion, which is actually the meaning of J, and of J^{-1}.

The viscous fluid is an example of dissipative system that, if it is *isolated* (i.e., without energetic supply from the environment), will relax to an *asymptotically stable field configuration*. Correspondingly, the total entropy of the isolated system (24) will grow up to a certain maximum value allowable S_{max}, as a *Lyapunov functional* is expected to do (Courbage & Prigogine, 1983).

The algebrization of the viscous fluid dynamics in LF will follow the same route proposed in Fish (2005) and references therein (see, in particular, Morrison (1984)), or in Materassi & Tassi (2012 b): the inspiration comes from the fact that, in the vicinity of the equilibrium, it is possible to represent the convergence of the system to the asymptotically stable point as a *metric system* (Morrison, 2009).

A metric system is a dynamical system of variables ψ that has a smooth function $S(\psi)$ on its phase space that generates the dynamics via a *symmetric, negative semidefinite* operator G, so that one has:

$$\dot{\psi}_i=-G_i^j\left(\psi\right)\frac{\partial S}{\partial\psi_j}. \tag{33}$$

It is possible, then, to define a symmetric Leibniz bracket (Guha, 2007) for two any observables f and g by stating:

$$\left(f,g\right)=-\frac{\partial f}{\partial\psi_i}G_i^j\left(\psi\right)\frac{\partial g}{\partial\psi_j}, \tag{34}$$

so that *the dynamics of the metric system* is prescribed as:

$$\dot{f}=\left(f,S\right). \tag{35}$$

It is possible to see that the asymptotically stable configuration will be a stationary point for $S(\psi)$, $\frac{\partial S}{\partial\psi_j}\ \Rightarrow\ \dot{\psi}_i=0$, and that it will be a maximum, since from (34) and (35) and the negative seminidefiniteness of G, the function $S(\psi)$ turns out to be monotonically growing with time:

$$\dot{S}=\left(S,S\right)=-\frac{\partial S}{\partial\psi_i}G_i^j\left(\psi\right)\frac{\partial S}{\partial\psi_j}\geq 0. \tag{36}$$

The smooth function $S(\psi)$ then appears to be *a Lyapunov function for the metric system.*

Since the metric system appears to be a good tool to describe the convergence to an asymptotically stable equilibrium, in order to let the fluid be described by a bracket algebra of observables that allows for relaxation, next to the Poisson bracket (21), one introduces a *metric bracket* (f, g), i.e.:

$$(f,g)_{\mathrm{LF}} = (g,f)_{\mathrm{LF}}, \quad (f,f)_{\mathrm{LF}} \leq 0 \quad \forall \quad f,g \in O_{\mathrm{fluid}}^{\mathrm{LF}}. \tag{37}$$

If the choice of the metric bracket is clever enough, i.e. provided the requirement

$$(H,f)_{\mathrm{LF}} = 0, \quad \forall \quad f \in O_{\mathrm{fluid}}^{\mathrm{LF}} \tag{38}$$

is satisfied, then the dissipative part of dynamics may be entrusted to this new algebraic object, with the choice of the total entropy $S[s]$ as the *metric generating function*, since we already know it is a Lyapunov functional of relaxation:

$$\dot{f}_{\mathrm{diss}} = \lambda (f,S)_{\mathrm{LF}}, \quad \forall \quad f \in O_{\mathrm{fluid}}^{\mathrm{LF}} \tag{39}$$

(this λ is a suitable constant that only makes a definite physical sense at the equilibrium).

The full dynamics may be obtained by putting together the *non-dissipative symplectic part* (I. 22) and the *dissipative metric part* (39), so to give rise to a properly algebrized dissipative system relaxing to some asymptotic equilibrium, namely a *metriplectic system* (Morrison, 1984; Guha, 2007):

$$\dot{f} = \{f,H\}_{\mathrm{LF}} + \lambda (f,S)_{\mathrm{LF}}, \quad \forall \quad f \in O_{\mathrm{fluid}}^{\mathrm{LF}}. \tag{40}$$

The Poisson bracket $\{.,.\}_{\mathrm{LF}}$ is the one defined in (21), while the metric component $(.,.)_{\mathrm{LF}}$ has been obtained in Materassi (2014) by reasoning on the application of the inverse map $\vec{\zeta} \mapsto \vec{a}$ (and of its implications in terms of functional derivatives) to the corresponding quantity presented in Morrison (1984) for the same system in the EF. The bracket reads:

$$\begin{aligned}(f,g)_{\mathrm{LF}} = \frac{1}{\lambda} \int_{D_0} J d^3 a \Bigg\{ & \kappa T^2 \nabla^{\eta} \left(\frac{1}{\rho_0 T} \frac{\delta f}{\delta s} \right) \nabla_{\eta} \left(\frac{1}{\rho_0 T} \frac{\delta g}{\delta s} \right) + \\ & + T \Lambda_{\alpha\beta\gamma\delta} \left[\nabla^{\alpha} \left(\frac{\delta f}{\delta \pi_{\beta}} \right) - \frac{1}{\rho_0 T} \nabla^{\alpha} \left(\frac{\pi^{\beta}}{\rho_0} \right) \frac{\delta f}{\delta s} \right] \left[\nabla^{\gamma} \left(\frac{\delta g}{\delta \pi_{\delta}} \right) - \frac{1}{\rho_0 T} \nabla^{\gamma} \left(\frac{\pi^{\delta}}{\rho_0} \right) \frac{\delta g}{\delta s} \right] \Bigg\}.\end{aligned} \tag{41}$$

http://dx.doi.org/10.5772/59319

It is possible to check that the application of the prescription (40) to the dynamical variables $\vec{\zeta}$, $\vec{\pi}$ and s reproduces the equations of motion for the dissipative fluids (32), as represented in the LF via parcel variables.

The former Poisson algebra of the observables of a fluid in the LF, the set $O_{\text{fluid}}^{\text{LF}}$, is now structured with both the brackets $\{.,.\}_{\text{LF}}$ and $(.,.)_{\text{LF}}$, so that a *metriplectic structure* may be defined by stating

$$\langle\langle f,g\rangle\rangle_{\text{LF}} = \{f,g\}_{\text{LF}} + (f,g)_{\text{LF}} \quad \forall \quad f,g \in O_{\text{fluid}}^{\text{LF}}. \tag{42}$$

In order for the metriplectic bracket $\langle\langle .,.\rangle\rangle_{\text{LF}}$ to generate a Leibniz dynamics (that corresponds to the dynamics of viscous fluids in the LF), the definition of the free energy

$$F\left[\vec{\zeta},\vec{\pi},s\right] = H\left[\vec{\zeta},\vec{\pi},s\right] + \lambda S\left[s\right] \tag{43}$$

is necessary. It is possible to see that, when the prescription

$$\dot{f} = \langle\langle f,F\rangle\rangle_{\text{LF}} \tag{44}$$

is applied to the dynamical variables ($\vec{\zeta}$, $\vec{\pi}$, s), the dynamics of the viscous fluid is re-obtained, as described by the equations of motion (32).

6. Conclusion

In this Chapter we have described the dynamics of viscous fluids in the formalism descending from the Lagrangian viewpoint, according to which the motion of the fluid parcels is followed along their evolution, just like it is done in discrete dynamics with the motion of point particles.

The "something different" that parcels have with respect to point particles is their internal structure, which is a "gas" (a fluid, truly) of microscopic particles, all in all described by their mass density, roughly J^{-1}, and their mass-specific entropy density s. The existence of this very rich internal structure has important consequences influencing the dynamics of the parcels moving through the space.

The particle ensemble may influence the parcel's motion both in a "conservative" and in a "dissipative" way.

On the one hand, particles may undergo "adiabatic interactions", hitting each other conserving their kinetic energy, and propagating, to the nearby particles, pressure: this is the effect of those motions of particles exerting forces, but not exchanging heat, i.e. not degrading the "ordered" energy of the macroscopic scales at which the parcel variables $\vec{\zeta}$ and $\vec{\pi}$ are defined. If the language of textbook Thermodynamics were to be used, we should state this is the kind of micro-macro interactions via which a parcel makes *mechanical work* on the nearby parcels, but does not give or receive heat.

These adiabatic interactions give rise to the pressure terms in (17) and (20), that still remain equations of motion of a dissipation-free system.

An ensemble of particles is however able to exchange *heat* too, i.e. transform the energy of macroscopic scales to micro-scale energy. This happens due to friction forces, that are representing in an effective way a huge amount of microscopic collisions in which particles literally drain kinetic energy from $\vec{\zeta}$ and $\vec{\pi}$ of nearby parcels. Or, the micro-motions and collisions among particles, from one parcel to the next one, give rise to *thermal conduction*, with a proper heat flux in which "nothing happens" at macroscopic scales, and nevertheless there is a transfer of internal energy throughout the fluid.

Two approaches are presented here to treat the dissipative phenomena of a viscous, thermally conducting fluid: writing the partial derivative equations of motion (32) in LF; or constructing the metriplectic algebra (42), through which the viscous, thermally conducting fluid is regarded as an isolated system with dissipation that has a way to relax to asymptotically stable equilibria.

In both approaches we have tried to stress the role of the *entropy functional* (24), that is fully enhanced in the metriplectic algebra (Materassi, 2014).

Let us conclude briefly by stressing again the limits of the LF as described here, i.e. via parcel variables. The fundamental requirements of parcel formulation in § 2, on which all the LF is based, render it possible to describe the evolution of fluid matter only in "rather regular" motions: no sub-fluid scale mixing, no evaporation, no droplet formation, no noisy micro-scale turbulence. Just the deformation of a body that can be regarded as an elastic rubber space-filling cloud, much more a shape shifting smooth Barbapapa rather than the water of a fountain dropping and fragmenting in pointy wavelets.

Author details

Massimo Materassi*

Address all correspondence to: massimo.materassi@fi.isc.cnr.it

Istituto dei Sistemi Complessi ISC-CNR, Italy

References

[1] Arnold, V. I. (1989). Mathematical Methods of Classical Mechanics (2nd ed.). New York: Springer, ISBN 978-0-387-96890-2.

[2] Bennett, A. F., Lagrangian Fluid Dynamics. *Monographs on Mechanics*, Cambridge University Press, Cambridge, (2006) ISBN-10: 0521853109.

[3] Chang, T. (1999). Self-organized criticality, multi-fractal spectra, sporadic localized reconnections and intermittent turbulence in the magnetotail, *Physics of Plasmas*, 6, 4137-4145.

[4] Courbage, M. & Prigogine, I. (1983). Intrinsic randomness and intrinsic irreversibility in classical dynamical systems, *Proceedings of the National Academy of Science of USA*, 80, pp. 2412-2416, Physics.

[5] Dewar, R.C., Lineweaver, C.H., Niven, R.K., Regenauer-Lieb, K. (editors) (2014). Beyond the Second Law-Entropy Production and Non-equilibrium Systems, Springer, ISBN 978-3-642-40154-1.

[6] Feynman, R. P. (1963) The Feynman Lectures on Physics, Online Edition, 1, § 10, "Conservation of Momentum", website: http://www.feynmanlectures.caltech.edu/I_10.html.

[7] Fish, D., Metriplectic Systems, Portland University (2005), PhD Thesis. Downloadable at: http://web.pdx.edu/~djf/.

[8] Goldstein, H. (2002). Classical Mechanics, Addison Wesley, ISBN 0-201-65702-3.

[9] Guha, P. (2007). Metriplectic structure, Leibniz dynamics and dissipative systems. *J. Math. Anal. Appl.* 326, 121.

[10] Lusanna, L., M. Materassi (2000). A Canonical Decomposition in Collective and Relative Variables of a Klein-Gordon Field in the Rest-Frame Wigner-Covariant Instant Form, *Int.J.Mod.Phys.* A15, 2821-2916, doi: 10.1142/S0217751X00001280.

[11] Materassi, M., Consolini, G. (2008). Turning the resistive MHD into a stochastic field theory, *Nonlinear Processes in Geophysics*, 15 (4), pp. 701-709.

[12] Materassi, M. and Tassi, E. (2012 a). Metriplectic Framework for Dissipative Magneto-Hydrodynamics, *Physica D*, 241, Issue 6, 729 (15 March 2012).

[13] Materassi & Tassi, E. (2012 b). Algebrizing friction: a brief look at the Metriplectic Formalism. *Intellectual Archive Journal*, Toronto: Shiny World (Issue of July 2012).

[14] Materassi, M., E. Tassi, G. Consolini (2012). Topics in Magnetohydrodynamics, Edited by Linjin Zheng, ISBN 978-953-51-0211-3, 220 pages, Publisher: InTech, Chapters published March 09, 2012 under CC BY 3.0 license.

[15] Materassi, M. (2014). Metriplectic Algebra for Dissipative Fluids in Lagrangian Formulation, submitted.

[16] Morrison, P.J.; (1984). Some Observations Regarding Brackets and Dissipation, *Center for Pure and Applied Mathematics Report PAM-228*, University of California, Berkeley.

[17] Morrison, P.J.; (2009). Thoughts on Brackets and Dissipation: Old and New, *Journal of Physics: Conference Series*, pp. 1-12, Vol. 169, 012006.

[18] Morrison, P. J. & Greene, J. M.; (1980). Noncanonical Hamiltonian Density Formulation of Hydrodynamics and Ideal Magnetohydrodynamics, *Physical Review Letters*, pp. 790-793, Vol. 45, No. 10, September 8.

[19] Padhye, N. S. (May 1998). Topics in Lagrangian and Hamiltonian fluid dynamics: Relabeling symmetry and ion-acoustic wave stability, dissertation presented to the Faculty of the Graduate School of The University of Texas at Austin in Partial Fullment of the Requirements for the Degree of PhD.

[20] Padhye, N. and P. J. Morrison (1996 a). Relabeling Symmetries in Hydrodynamics and Magnetohydrodynamics, *Plasma Physics Reports* 22, 869.

[21] Padhye, N. and P. J. Morrison (1996 b). Fluid Element Relabeling Symmetry, *Physics Letters* 219A, 287.

[22] Rai Choudhuri, A. (1998). The Physics of Fluids and Plasmas-an introduction for astrophysicists, Cambridge University Press.

[23] Zemansky, M. W. and Dittman, R. H. (1996). Heat and Thermodynamics, Mcgraw-Hill College; 7th edition.

Chapter 6

Evolution of Small Perturbations of the Free Surface of Viscous Fluid in the Stokes Approximation

Tamara Pogosian, Ivan Melikhov,
Anastasia Shutova, Alexey Amosov and
Sergey Chivilikhin

Additional information is available at the end of the chapter

http://dx.doi.org/10.5772/59226

1. Introduction

Stokes steady-state approximation [1] is widely used for calculation of viscous flows with low Reynolds numbers. This approach was first applied to describe the flow with free boundaries in [2]. Many classical problems of high-viscous liquids hydrodynamics were solved in this approximation: the destruction of bubbles [3], the formation of the cusp on the liquid surface [4,5], the coalescence of liquid particles [6] and some problems of nanofluidics [7].

The idea of using inertia-less approximation to calculate wave relaxation on the surface of a fluid with high viscosity was suggested by [8]. The relaxation of harmonic perturbations in this approximation was considered by [9,10]. This approach has also been used in studying thin film flow [11], two-dimensional Stokes flow with free boundary [12-13].

In this work we study an aperiodic relaxation of a small localized perturbation of the planar surface of an incompressible fluid under the influence of gravity and surface tension. The description is given in the Stokes approximation. It is shown that this imposes limitations on the three-dimensional spectrum of the considered perturbations. An equation describing perturbation damping in Fourier space was obtained. It is shown that the volume of the perturbation becomes zero in the time that is small by comparison to the time of perturbation damping. In the short wave limit, the law of evolution of the perturbation permits a simple geometric interpretation.

The second part of the work is focused on theoretical investigation of instability of the surface shape under the influence of diffusion mass flux through the surface. A local eminence on the surface can lead to perturbation growth. Stability and instability of the perturbation will depend on which mechanism prevails. Diffusion causes distortion of surface shape, gravitational and capillary forces have the opposite effect.

2. Evolution of half-space free boundary perturbation

2.1. The general equations

We consider a half-space filled with a highly viscous incompressible Newton liquid of constant viscosity. At the initial moment of time the planar surface of half-space obtains a small local perturbation. We will investigate further the evolution of this perturbation due to gravitation, capillary forces and external mass flux through the free surface. We introduce the Cartesian coordinate system with axes (x_1, x_2) located on the unperturbed surface and the axis z directed along the external normal to it. Then the perturbation of the surface can be defined as $z=h(x, t)$, where x is the vector with components x_1, x_2 and t is the time. We express the equations of motion in the Stokes approximation [1], the continuity equation, the equation for the renormalized pressure $\Pi=P+\rho gz$ (where P is the true pressure) as

$$\begin{cases} \partial_{zz}^2 V_z + \partial_{\beta\beta}^2 V_z = \frac{1}{\mu}\partial_z \Pi, \\ \partial_{zz}^2 V_\alpha + \partial_{\beta\beta}^2 V_\alpha = \frac{1}{\mu}\partial_\alpha \Pi, \\ \partial_z V_z + \partial_\beta V_\beta = 0, \\ \partial_{zz}^2 \Pi + \partial_{\beta\beta}^2 \Pi = 0. \end{cases} \qquad (1)$$

The indices α, β can be 1 or 2; summation over repeated indices is assumed. The boundary conditions which express the continuity of the normal and tangential stresses on the free surface, in the approximation linear by h, have the form

$$\left(-\Pi + 2\mu\partial_z V_z\right)\Big|_{z=0} = \sigma\partial_{\beta\beta}^2 h - \rho g h, \quad \left(\partial_\alpha V_z + \partial_z V_\alpha\right)\Big|_{z=0} = 0. \qquad (2)$$

The following conditions in case of infinite half-space must also be met:

$$\left.\begin{pmatrix} V_z \\ V_\alpha \\ \Pi \end{pmatrix}\right|_{z\to-\infty} = 0. \qquad (3)$$

Here V_z and V_α are the components of velocity, ∂_z and ∂_α are the derivatives along the coordinates z and x_α accordingly, g is the acceleration due to gravity; μ, ρ and σ are the viscosity, density and surface tension coefficient of the liquid.

Equation of free surface evolution and its initial condition have the form

$$\partial_t h = V_z\big|_{z=0} + W, \quad h\big|_{t=0} = h^0(\mathbf{x}), \tag{4}$$

where W describes the influence of external mass fluxes on the velocity of free boundary.

Applying to (1)-(4) the two-dimensional Fourier transform with respect to the longitudinal coordinates

$$f_{\mathbf{k}} = \int f(\mathbf{x}) e^{-i\mathbf{k}\mathbf{x}} d^2x$$

we obtain

$$\begin{cases} \partial_{zz}^2 V_{z\mathbf{k}} - k^2 V_{z\mathbf{k}} = \dfrac{1}{\mu}\partial_z \Pi_{\mathbf{k}} \\ \partial_{zz}^2 V_{\alpha\mathbf{k}} - k^2 V_{\alpha\mathbf{k}} = \dfrac{ik_\alpha}{\mu}\Pi_{\mathbf{k}} \\ \partial_z V_{z\mathbf{k}} + ik_\beta V_{\beta\mathbf{k}} = 0 \\ \partial_{zz}^2 \Pi_{\mathbf{k}} - k^2 \Pi_{\mathbf{k}} = 0 \end{cases} \tag{5}$$

$$\left(-\Pi_{\mathbf{k}} + 2\mu\partial_z V_{z\mathbf{k}}\right)\big|_{z=0} = -\left(\sigma k^2 + \rho g\right)h_{\mathbf{k}}, \quad \left(ik_\alpha V_{z\mathbf{k}} + \partial_z V_{\alpha\mathbf{k}}\right)\big|_{z=0} = 0, \quad \left.\begin{pmatrix} V_{z\mathbf{k}} \\ V_{\alpha\mathbf{k}} \\ \Pi_{\mathbf{k}} \end{pmatrix}\right|_{z\to-\infty} = 0, \tag{6}$$

$$\partial_t h_{\mathbf{k}} = V_{z\mathbf{k}}\big|_{x_\perp=0} + W_{\mathbf{k}}. \tag{7}$$

Integrating (5) with boundary conditions (6), we obtain

$$\begin{pmatrix} V_{z\mathbf{k}} \\ V_{\alpha\mathbf{k}} \\ \Pi_{\mathbf{k}} \end{pmatrix} = \begin{pmatrix} kz - 1 \\ ik_\alpha z \\ 2\mu k \end{pmatrix} \frac{\sigma k^2 + \rho g}{2\mu k} h_{\mathbf{k}} e^{kz}. \tag{8}$$

It is apparent that the Fourier component $h_{\mathbf{k}}$ has an effect in the layer of depth k^{-1}. Based on (7), (8), the equation of the free surface motion can be written as

$$\partial_t h_{\mathbf{k}} + \frac{1}{2\mu k}\left(\sigma k^2 + \rho g\right)h_{\mathbf{k}} = W_{\mathbf{k}} \tag{9}$$

2.2. Relaxation of the initial perturbation of the liquid surface, exact solution

Consider the relaxation of the initial perturbation of the surface due to capillary and gravitation forces in the absence of mass flux at the boundary ($W_k = 0$). In this case, the solution of equation (9) with initial condition $h_k|_{t=0} = h_k^0$ becomes

$$h_k = h_k^0 e^{-t/T(k)}, \tag{10}$$

where $T(k) = \dfrac{2\mu k}{\sigma k^2 + \rho g}$ is the characteristic relaxation time of the corresponding harmonics. For $k \to 0$ and $k \to \infty$ the relaxation time tends to zero assuming non-zero gravity and surface tension, respectively. At $k = k_* = (\rho g / \sigma)^{1/2}$ the relaxation time has its maximal value $T_* = \mu / (\rho g \sigma)^{1/2}$. Small relaxation time for high-frequency harmonics means that the surface tension forces lead to a rapid smoothing of sharp irregularities. On the other hand $T|_{k=0} = 0$, therefore $h_k|_{k=0} = 0$ at $t > 0$. Noting that

$$h_k|_{k=0} = \int h(\mathbf{x}, t)\, d^2x$$

constitutes the "volume of perturbation". According to (10) this value becomes equal to zero due to the gravity force. Immediacy of this process is determined using a quasi-stationary Stokes approximation. Detailed discussion of this topic see in [14].

Switching to the case of arbitrary wave numbers, we introduce the dimensionless variables

$$\tau = t/T_*, \quad \theta = T/T_*, \quad \xi = k_* \mathbf{x}, \quad \boldsymbol{\kappa} = \mathbf{k} / k_*.$$

Then (10) takes the form

$$h_\kappa = h_\kappa^0 e^{-\tau/\theta(\kappa)}, \quad \theta(\kappa) = \frac{2\kappa}{\kappa^2 + 1}, \quad h_\kappa = \int h(\xi, \tau) e^{-i\kappa\xi}. \tag{11}$$

Applying inverse Fourier transform to (11) we have

$$h(\xi, \tau) = \int g(\xi - \xi', \tau) h^0(\xi') d^2\xi', \quad g(\xi, \tau) = \int e^{i\kappa\xi - \tau/\theta(\kappa)} \frac{d^2\kappa}{(2\pi)^2}. \tag{12}$$

The expression (12) allows calculation of the surface perturbation in any time for an arbitrary initial perturbation.

2.3. Qualitative description of the perturbation relaxation

For a qualitative description of the relaxation of the initial perturbation is convenient to use the original expression (11). We have

$$h(\xi,\tau)=\int e^{i\kappa\xi}h_{\kappa}^{0}e^{-\tau/\theta(\kappa)}\frac{d^2\kappa}{(2\pi)^2},\quad \theta(\kappa)=\frac{2\kappa}{k^2+1}. \tag{13}$$

From (13) it is clear that the configuration at the moment τ is determined by the Fourier transform of the initial perturbation h_{κ}^{0} and the multiplier $e^{-\tau/\theta(\kappa)}$. The maximum value of this factor is equal to $e^{-\tau}$, and corresponds to $\tau=1$. Let $\kappa_{1,2}(\tau)$ be the values of κ at which the multiplier takes a value in e times less than the maximum-see Fig. 1.

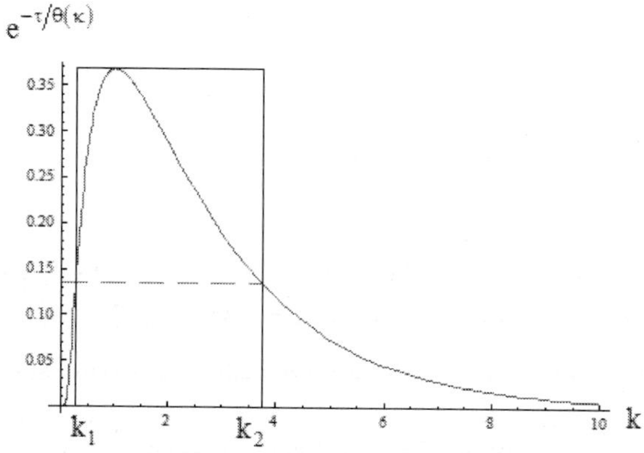

Figure 1. Graph of the function $e^{-\tau/\theta(\kappa)}$ at $\tau=1$ and its approximation with a step function.

By solving the equation $\frac{\tau}{\theta(\kappa)}=\tau+1$, we obtain

$$\kappa_{1,2}=\frac{1}{\tau}\left(\tau+1\pm\sqrt{2\tau+1}\right). \tag{14}$$

At $\tau\to 0$, $\kappa_1=\frac{\tau}{2}$, $\kappa_2=\frac{2}{\tau}$. On the other hand, at $\tau\to\infty$, $\kappa_{1,2}=1\mp\sqrt{\frac{2}{\tau}}$.

To estimate perturbation evolution we replace function $e^{-\tau/\theta(\kappa)}$ with the step function $f(\tau,\kappa)$ which is equal to zero for $\kappa<\kappa_1(\tau)$ and for $\kappa>\kappa_2(\tau)$. In the interval $\kappa_1(\tau)\le\kappa\le\kappa_2(\tau)$ this function is equal to $e^{-\tau}$-see Fig.1:

$$f(\kappa,\tau) = e^{-\tau}\vartheta(\kappa - \kappa_1(\tau))\vartheta(\kappa_2(\tau) - \kappa), \tag{15}$$

where $\vartheta(x)$ is the Heaviside step function.

Returning to the expression (13) we see that the evolution of perturbations can be considered as a result the initial perturbation "cut-off" in Fourier space. The system under consideration behaves as filter which "cuts" in the spectrum of the initial perturbation wave numbers the annular region with characteristic dimensions $\kappa_1(\tau) \le |\boldsymbol{\kappa}| \le \kappa_2(\tau)$. At the same time the amplitude of the Fourier transform of the perturbation decreases as $e^{-\tau}$. It can be seen that the width of the annular region decreases with increasing time, rapidly at the beginning-mainly due to the reduction of the outer radius of the ring κ_2, then slowly-due to symmetric increasing κ_1 and decreasing κ_2. At large times, the ring is localized in the neighborhood of a circle of radius $\kappa = 1$ and has a characteristic width $\Delta\kappa = 2\sqrt{\frac{2}{\tau}}$. Such filtering process determines the qualitative features of the relaxation of the initial perturbation.

Spatial filtering of the Fourier transform of the initial perturbation in the region of small wave numbers is equivalent to subtracting its part localized in a circle of radius κ_1. It leads to the subtraction of additional perturbation localized in a circle with radius $\xi_1 = \frac{2\pi}{k_1}$ from the initial perturbation. When $\tau \to 0$, $\xi_1 = \frac{4\pi}{\tau} \to \infty$. This long-range effect of the initial perturbation is determined by the force of gravity. This is clearly seen in the transition to dimensional movements: the radius of the long-range perturbations is $r_1 = \frac{\xi_1}{k_*} = \frac{4\pi\mu}{\rho g t}$. Figuratively speaking, the perturbation is drowning in the ambient fluid, distorting the shape of the surface at long distance.

Localization of the Fourier transform of the perturbation in a circle of radius κ_2 leads to an expansion of its external borders, in the coordinate representation, to a circle of radius $\xi_2 = \frac{2\pi}{\kappa_2}$. In dimensional variables $r_2 = \frac{\xi_2}{k_*} = \frac{\pi\sigma t}{\mu}$, i.e. this effect is due to the action of surface tension forces.

To evaluate the value of the perturbation at the origin of coordinates we use of the fact that

$$h(0,\ \tau) = \int h_\kappa(\tau)\frac{d^2\kappa}{(2\pi)^2}$$

Then

$$h(0,\ \tau) = \int\limits_{\kappa_1(\tau)\le|\kappa|\le\kappa_2(\tau)} h_\kappa^0 \frac{d^2\kappa}{(2\pi)^2}$$

The above statements are based on the approximate replacement of real $e^{-\tau/\theta(\kappa)}$ functions with the step function (15). Therefore, we can only expect an approximate correspondence between the description, based on these statements, and the actual process of the perturbation relaxation.

Let us consider the relaxation of perturbations with different spatial scale and shape. To illustrate the qualitative dependencies we will use graphics produced on the basis of the exact solution (12).

2.4. Relaxation long-wave perturbations by gravity

If the characteristic dimensionless scale of the region occupied by the perturbation $a >> 1$, then the Fourier transform of this disturbance is concentrated in the region with a typical size $\frac{2\pi}{a} << 1$. At the initial stage the relaxation of such perturbation is mainly determined by filtering out small wave numbers of its Fourier representation, i.e., by gravity. In this case, as shown above, to the initial perturbation is added, taken with the opposite sign, the Fourier inverse transform of the initial perturbation, lying in a circle of radius $\kappa_1(\tau)$. In the coordinate representation this additive is concentrated in a circle of radius $\xi_1(\tau) = 2\pi / \kappa_1(\tau)$ and has an amplitude

$$\Delta h(\tau) = \int_{|\kappa| \le \kappa_1(\tau)} h_\kappa^0 \frac{d^2\kappa}{(2\pi)^2}$$

At the initial stage of relaxation process $\kappa_1 << 2\pi / a$, respectively, $\xi_1 >> a$,

$$\Delta h(\tau) = \frac{1}{4\pi} \kappa_1^2(\tau) h_\kappa^0 \big|_{\kappa=0} = \frac{\tau^2}{16\pi} \int h^0(\xi)\, d^2\xi$$

Through this long-wavelength addition, the initial perturbation goes down (in case of positive volume) on $\Delta h(\tau)$, practically without distortion of the shape. Outside the region of the initial perturbation arises an additional perturbation with the amplitude $\Delta h(\tau)$ and attenuation radius $\xi_1(\tau)$. At large times, the shape of additional perturbation is somewhat distorted, its radius decreases and becomes comparable to the size of the initial perturbation.

By the time $\tau_1 = 4\pi / a$, when $\kappa_1 = 2\pi / a$, the process of relaxation of the perturbation is mainly completed. To illustrate these effects on the Fig.2 presented the relaxation of axisymmetric long-wave perturbation shape. On the Fig.3 we can see the increasing of this perturbation amplitude.

If the characteristic dimensions of the initial perturbation in two mutually perpendicular directions are very different, the relaxation process is more complicated. Note that for the perturbation, elongated along an axis ξ_1 ($a_1 >> a_2 >> 1$), the region of localization of the Fourier transform has the inverse aspect ratio $2\pi / a_1 << 2\pi / a_2$. Removal from the spectrum of the initial perturbation the circle with radius $\kappa_1(\tau) = \tau / 2$ leads initially to the formation of a symmetric long-wave perturbation of the radius $4\pi/\tau$ and the amplitude $\Delta h = \frac{\tau^2}{16\pi}\int h^0(\xi)d^2\xi$. At this stage, the difference of the transverse scales the perturbation does not affect practically the nature of its evolution. Perturbation settles as a whole, and its amplitude decreases according to $h(0,\tau)=h^0(0)-\Delta h(\tau)$.

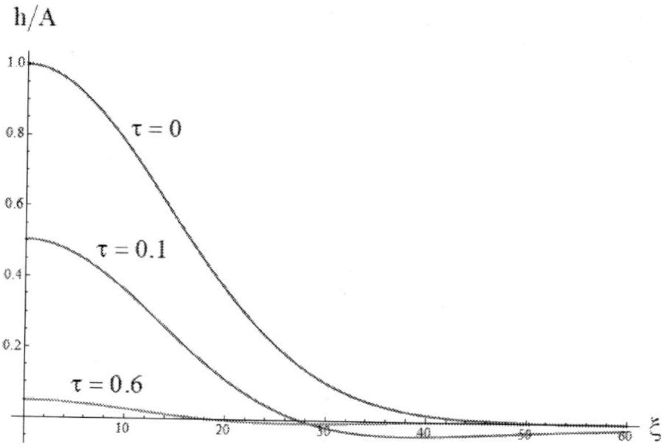

Figure 2. The relaxation of axisymmetric long-wave perturbations. The initial shape of perturbation is $h^0 = Ae^{-\xi^2/a^2}$, $a=20$.

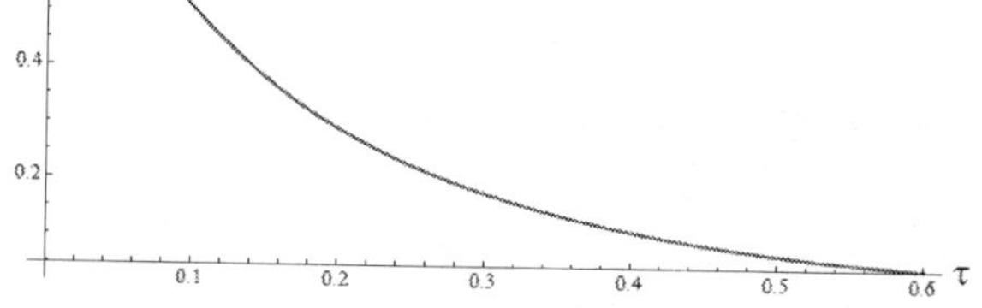

Figure 3. Increasing of amplitude of long-wave axisymmetric perturbation. The initial shape of perturbation is $h^0 = Ae^{-\xi^2/a^2}$, $a=20$.

http://dx.doi.org/10.5772/59226

From the time moment $\tau = 4\pi / a_1$, when $k_1(\tau) = 4\pi / a_1$, the area of influence of long-range perturbations loses circular symmetry. In the direction of the axis ξ_1 the scale of the area ξ_1^* becomes, and subsequently maintained, equal to a_1, and along the axis ξ_2 this area continues to shrink as $\xi_2^* = 4\pi / \tau$. The amplitude of the perturbation begins to decrease linearly $h(0, \tau) = h^0(0) - \Delta h(\tau)$, $\Delta h = \frac{\tau}{4\pi L_1} \int h^0(\xi) d^2\xi$.

At the moment $\tau = 4\pi / a_2$, when $\kappa_1(\tau) = 2\pi / a_2$ the perturbation is practically relaxed. Figure 4 shows the characteristic configuration of the long-wavelength perturbation, the extent of which the two axes are substantially different.

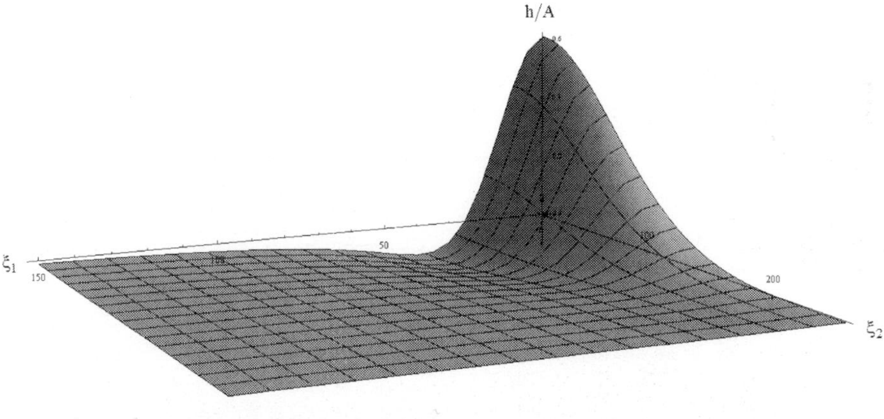

Figure 4. The configuration of the long-wave non-axisymmetric perturbations in the time $\tau = 0.04$. The initial shape of perturbation is $h^0 = A\exp\left(-\frac{\xi_1^2}{a_1^2} - \frac{\xi_2^2}{a_2^2}\right)$, $a_1 = 100$, $a_2 = 20$.

2.5. Relaxation of short-wave perturbations due to capillary forces

The evolution of perturbation with the scale $a << 1$ is different. The Fourier transform of this disturbance is concentrated in the region with a typical size of $2\pi / a >> 1$. In this case occurs the evolution of the perturbation in the initial stage due to filtration of its Fourier transform through small wave numbers. However, the contribution of this region to the total amplitude of the perturbation is small, and therefore, the effects of subsidence of the perturbation and the formation of the region of the far influence, in this case, is irrelevant.

Relaxation of short-wave perturbations is mainly determined by the action of surface tension forces, i.e. filtration its Fourier transform by the large wave numbers. The relaxation process begins at the moment $\tau = a / \pi$, when $\kappa_2 = 2\pi / a$. The size of the region occupied by the perturbation begins to grow as $\xi(\tau) = 2\pi / \kappa_2 = \pi\tau$, and its amplitude decreases at the initial stage, as $\frac{1}{\pi\tau^2}\int h^0(\xi)d^2\xi$. Characteristic decay time of the perturbation is of order unity. Fig. 5 shows the relaxation of short-wave axisymmetric perturbation.

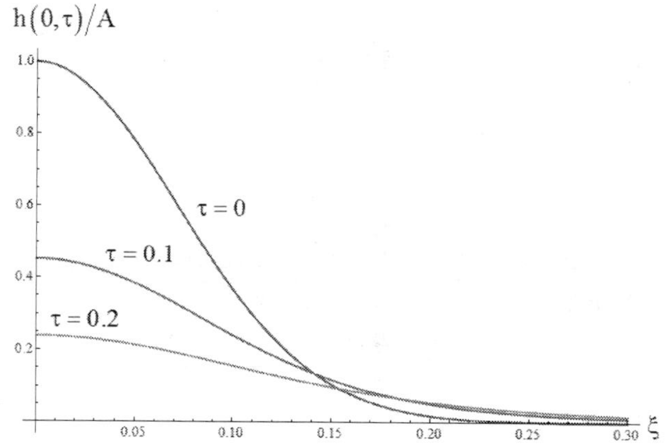

Figure 5. Relaxation of short-wave axisymmetric perturbation. The initial shape of perturbation $h^0 = Ae^{-\xi^2/a^2}$, $a = 0.1$.

Finally let's describe the relaxation of small-scale perturbations with significantly different scales along the axes ξ_1 and ξ_2 : $a_1 << a_2 << 1$. The evolution of this perturbation begins at the moment $\tau = a_1 / \pi$. The scale begins to increase like $\pi\tau$ in the ξ_1 axis direction and in ξ_2 axis direction it varies slightly. The amplitude of the perturbation decreases according to $h(0, \tau) = \frac{2}{\pi a_2 \tau}\int h^0(\xi)d^2\xi$.

After the moment $\tau = a_2 / \pi$ the magnitudes of the perturbation in the direction of the two axes become comparable. From this moment the perturbation becomes axially symmetric, and its radius increases as $\pi\tau$. Amplitude of perturbation thus decreases faster than during the first stage: $h(0, \tau) = \frac{1}{\pi\tau^2}\int h^0(\xi)d^2\xi$.

The relaxation process ends substantially at the moment $\tau = 1$. Fig.5 shows relaxation of short-wave non-axisymmetric perturbations. One can see that over time the perturbation becomes axisymmetric.

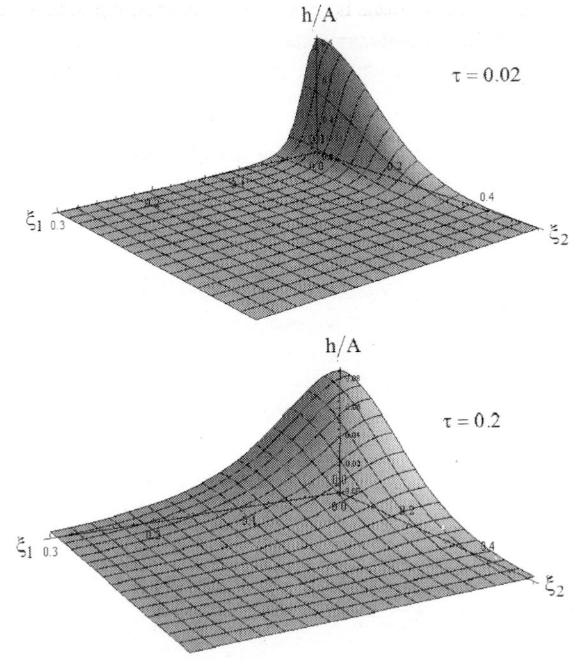

Figure 6. Configurations of the short-wave non-axisymmetric perturbation. The initial shape of perturbation is $h^0 = A\exp\left(-\frac{\xi_1^2}{a_1^2} - \frac{\xi_2^2}{a_2^2}\right)$, $a_1 = 0.02$, $a_2 = 0.2$.

3. Finite layer over solid bottom and over a half-space

3.1. Finite layer over solid bottom

In this section we study surface perturbations of a flat liquid layer with depth H resting on a solid bottom. We use system (1) and conditions on the free boundary (2) to calculate perturbation evolution due to gravity and surface tension. The conditions at infinity (3) are replaced by the no-slip condition along the solid surface

$$V_z\big|_{z=-H} = 0, \quad V_\alpha\big|_{z=-H} = 0. \tag{16}$$

Again applying the Fourier transform in the longitudinal coordinates and solving the resulting system of ordinary differential equations, we obtain

$$V_{zk} = \left(V_{zk0} - \frac{H\Pi_{k0}}{2\mu} \right) \frac{sh(k(z+H))}{sh(kH)} + \frac{(z+H)\Pi_k}{2\mu},$$
$$\Pi_k = \Pi_{k0} \frac{sh(k(z+H))}{sh(kH)} - \Pi_{kH} \frac{sh(kz)}{sh(kH)}, \tag{17}$$

where

$$\begin{pmatrix} V_{zk0} \\ \Pi_{k0} \\ \Pi_{kH} \end{pmatrix} = \begin{pmatrix} 2kH - sh(2kh) \\ 2\mu k(ch(2kH)+1) \\ 4\mu k(kHsh(kH)+ch(kH)) \end{pmatrix} \frac{\sigma k^2 + \rho g}{2\mu k\left(ch(2kH)+1+2(kH)^2\right)} h_k. \tag{18}$$

The evolution equation of the free surface and the initial condition has the form

$$\partial_t h_k + \frac{2kH - sh(2kh)}{ch(2kH)+1+2(kH)^2} \frac{\sigma k^2 + \rho g}{2\mu k} h_k = 0, \quad h_k|_{t=0} = h_k^0. \tag{19}$$

The solution of equation (19) is similar to (10)

$$h_k = h_k^0 e^{-t/T(k)}, \tag{20}$$

where $T(k) = \frac{ch(2kH)+1+2(kH)^2}{sh(2kh)-2kH} \frac{2\mu k}{\sigma k^2 + \rho g}$ is the characteristic relaxation time of corresponding harmonic.

At $kH >> 1$, $T(k, H) = \frac{2\mu k}{\sigma k^2 + \rho g}$ coincides with (10), i.e. with the case of a fluid of infinite depth. Therefore, for $k \to \infty$ the relaxation time $T(H, k) \to 0$ due to the surface tension. On the other hand, for $k \to 0$, $T(H, k) \to \infty$ unlike the case of a fluid of infinite depth. Then, according to (20),

$h_k |_{k=0} = h_k^0 |_{k=0}$ or $\int h(x, t) d^2x = \int h^0(x) d^2x$, i.e. the "volume of perturbation" is kept. Using the same scales as we used for infinite liquid, we present (20) in dimensionless form

$$h_\kappa = h_\kappa^0 e^{-\tau/\theta(\kappa,\bar{H})}, \quad \theta(\kappa, \bar{H}) = \frac{ch(2\kappa\bar{H})+1+2(\kappa\bar{H})^2}{sh(2\kappa\bar{H})-2\kappa\bar{H}} \frac{2\kappa}{\kappa^2+1}, \quad h_\kappa = \int h(\xi, \tau) e^{-i\kappa\xi}, \tag{21}$$

where $\bar{H} = k_* H$ is the dimensionless thickness of layer.

3.2. Finite layer over a half-space

Consider a plane layer with thickness H, filled with liquid with viscosity μ_1. Layer is located over the half-space which is filled with liquid viscosity μ_2. At the interface between the layer

http://dx.doi.org/10.5772/59226

and half-space we require continuity of the velocity components and continuity of the convolution of the stress tensor with the normal vector. Taking into account the gravitational forces and capillary forces on the free surface of the layer, using the technique described above, we obtain the equation of evolution of the free surface of the layer:

$$h_\kappa = h_\kappa^0 e^{-\tau/\theta(\kappa,\bar{H})}, \quad \theta(\kappa,\bar{H},m) = \frac{2m\cdot \mathrm{sh}(2\kappa\bar{H}) + (m^2+1)\mathrm{ch}(2\kappa\bar{H}) + (1+2\kappa^2\bar{H}^2)(m^2-1)}{(m^2+1)\mathrm{sh}(2\kappa\bar{H}) + 2m\cdot \mathrm{ch}(2\kappa\bar{H}) - 2\kappa\bar{H}(m^2-1)} \frac{2\kappa}{\kappa^2+1}, \quad (22)$$

where $m = \frac{\mu_2}{\mu_1}$. As expected, the characteristic relaxation time $\theta(\kappa, \bar{H}, m)$ at $m=1$ or $\bar{H} \to \infty$ transferred to the corresponding expression (13) obtained for the half-space, at $m \to \infty$, (22) transferred to (21)-expression for finite layer over solid bottom. At $\kappa=0$, $\theta(\kappa, \bar{H}, m)=0$. Consequently, the "volume of perturbation" instantly vanishes, as in the case of the half-space.

3.3. Comparing the cases of finite layer and finite layer over a half-space

Fig.7 shows function $e^{-\tau/\theta}$ for the cases of finite layer and finite layer over a half-space. Qualitative differences in the behavior of this function lead to a difference in the perturbation decay for these cases.

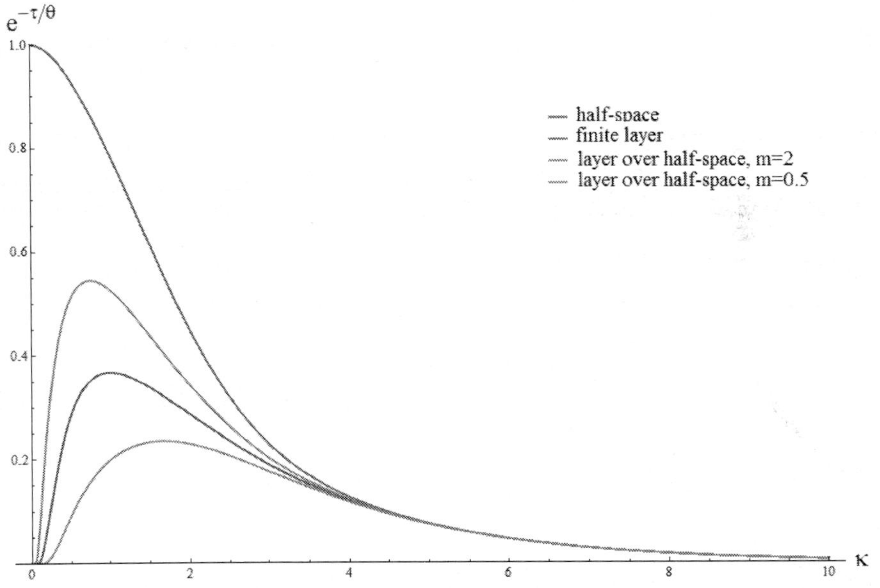

Figure 7. Graph of the function $e^{-\tau/\theta}$ at $\tau=1$ for half-space, finite layer and layer with half-space.

Fig. 8 and Fig. 9 show decrease of the amplitude for the perturbation with the same initial characteristics. The figures show that the perturbation on the surface layer over solid bottom decays faster than the perturbation on the surface layer over the half-space.

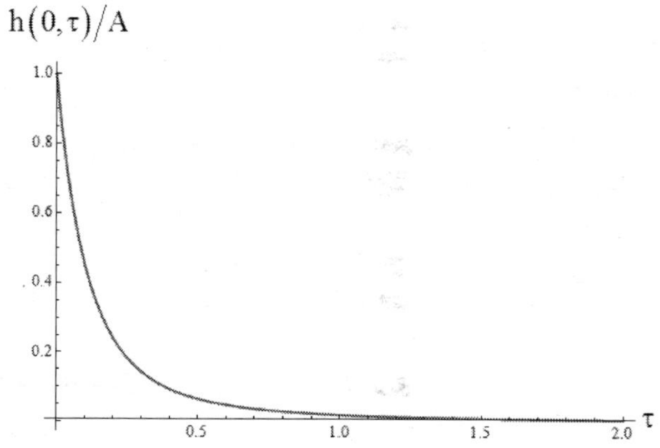

Figure 8. Relaxation of axisymmetric perturbation of free surface of the layer with thickness $\overline{H}=1$.The initial shape of perturbation $h^0 = Ae^{-\xi^2/a^2}$, $a=0.1$

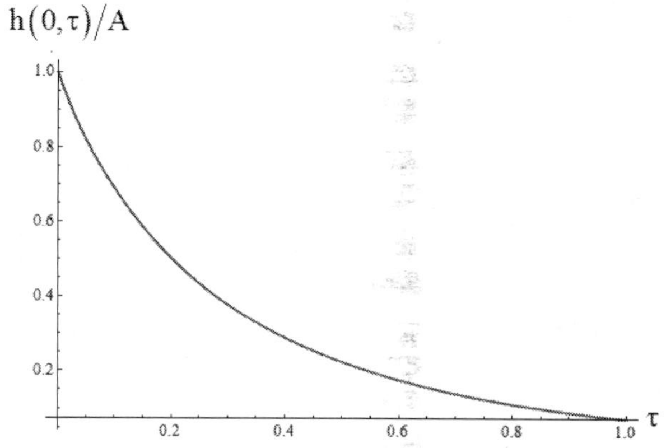

Figure 9. Relaxation of axisymmetric perturbation of free surface of the layer with thickness $\overline{H}=1$ over a half-space. The initial shape of perturbation $h^0 = Ae^{-\xi^2/a^2}$, $a=0.1$, $m=2$.

3.4. Comparing the relaxation of finite layer free surface perturbation in linear and non-linear approximation

Using finite-element software Comsol Multiphysics we developed a 2D axisymmetric model of the viscous flow to calculate transient relaxation of the perturbation on the free surface and compare it with analytical results reported in the previous sections. We used "Laminar Two-Phase Flow, Moving Mesh" module to enable deformation of the computational domain during the solution. Full Navier-Stokes equations are solved on the moving mesh to describe deformation of the domain due to capillary forces acting along the free surface.

Computational mesh was refined near the perturbation to resolve high curvature and velocity gradients (see Fig. 10).

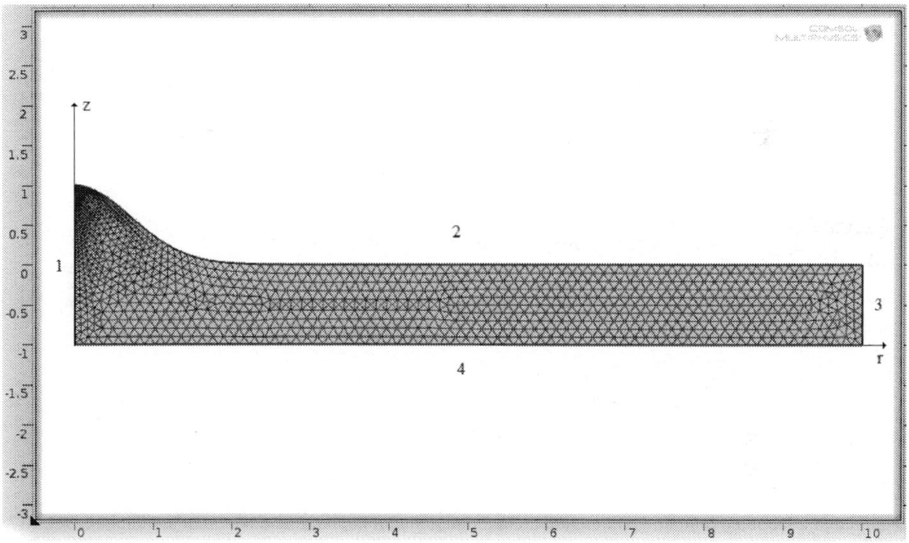

Figure 10. Initial non-deformed finite-element discretization of the domain. The initial shape of perturbation $h^0 = Ae^{-\xi^2/a^2}$, $A=1$, $a=1$. Boundary conditions: 1 – axial symmetry, 2 – free surface, 3 – slip, and 4 – no-slip wall.

Multiple cases were calculated corresponding to small and large non-dimensional perturbation amplitude and width. The evolution of the perturbation amplitude with time was compared for numerical and analytical models (Fig. 11). It can be seen that the analytics and FEM are in excellent agreement for small amplitude A=0.1 for both small and large width (a=0.1, a=1.0). Some discrepancy is observed in the case of large amplitude A=1.0 due to violation of the analytic model assumptions.

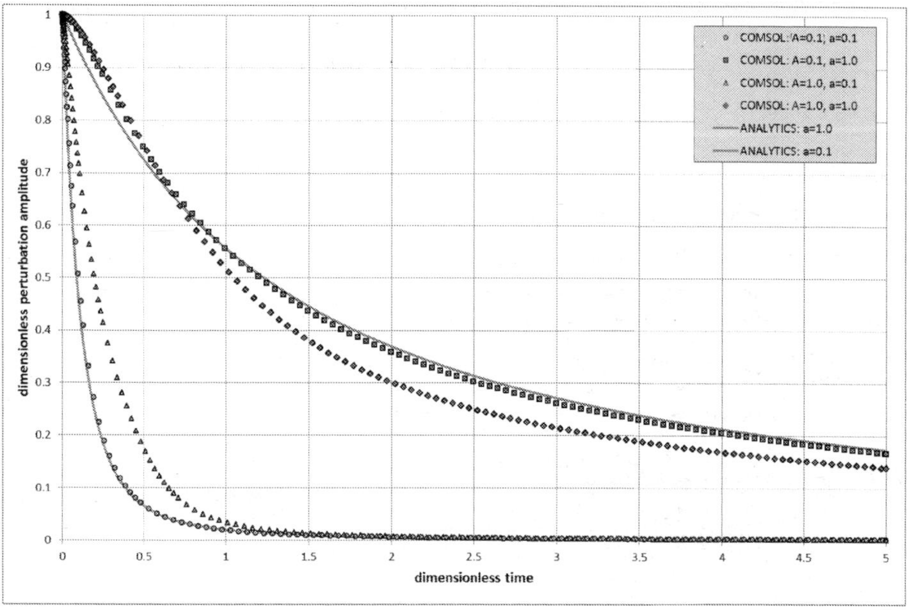

Figure 11. Comparison of the analytical model (solid lines) and FEM (points) results

4. Evolution of small perturbations of half-space free boundary due to external mass flux, gravitation and capillary forces

Let's describe the influence of external mass flow on stability of the free surface of half-space. A small perturbation on the free surface leads to a perturbation of the diffusion boundary layer above it. The corresponding mass flow perturbations on the free boundary defines the source term in the right hand side of (9) $W_k = \kappa V cth(kH_D)$, where H_D is the thickness of the diffusion boundary layer, V is the unperturbed velocity of the liquid level rise due to external mass flow. In this case the equation of perturbation evolution has the following shape

$$\partial_t h_k + \gamma h_k = 0, \quad h_k = h_k^0 e^{\gamma(k)t}, \tag{23}$$

where $\gamma(k) = -\frac{1}{2\mu k}(\sigma k^2 + \rho g) + kVcth(kH_D)$. At $\gamma > 0$ correspondent harmonic is unstable, and at $\gamma < 0$ it is stable. On the Fig.12 we can see the regions in the space of parameters, correspondent stable and unstable regimes. On the Fig.13 present the graphs of $\gamma(k)$ for these regions. In the case (a) all the harmonics with the wave number exceeding some critical value are unstable. In the case (b) all harmonics are stable. In the case (c) there is a range of wave numbers corresponding to unstable harmonics. In the case (d) the harmonics in a certain range of wave numbers, and with a wave number exceeding a critical value are unstable.

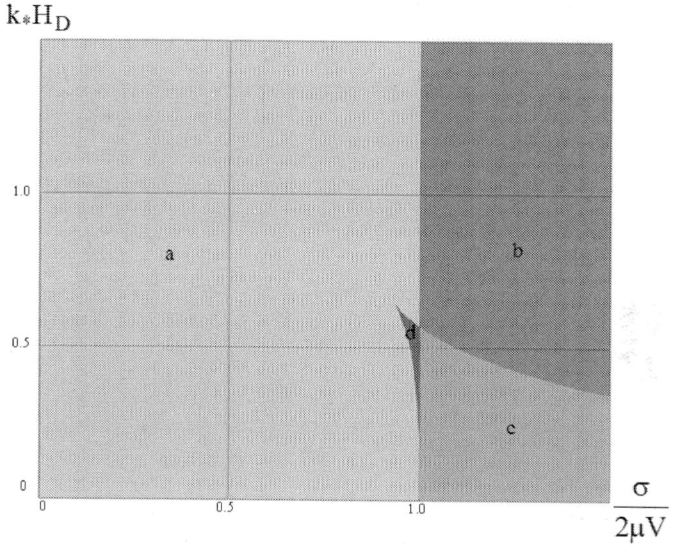

Figure 12. Space of parameters with regions, corresponds to regimes a-unstable, b-stable, c-partly unstable, d-mixed.

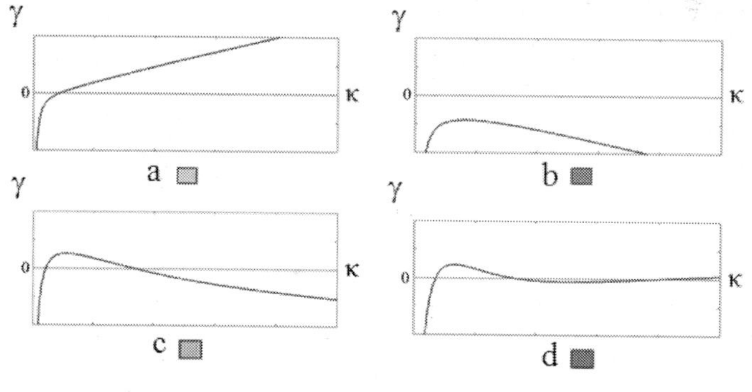

Figure 13. Graphs of $\gamma(k)$ for regimes: a-unstable, b-stable, c-partly unstable, d-mixed

5. Conclusion

The relaxation process of a small perturbation of the free planar surface of a viscous incompressible fluid under the action of capillary and gravitational forces was investigated. The case of half-space, layer over hard bottom and layer over the half-space was analyzed. We propose a method for qualitative analysis of the configuration of the perturbation based on spatial filtering the Fourier transform of the initial perturbation.

It is shown that under the gravity force there is an additional long-wave perturbation arising around the initial perturbation which provides a zero total volume of the perturbation. Capillary forces lead to an increase of the area affected by the perturbation. The influence of amplitude of perturbation on the rate of its relaxation was investigated numerically and compared with the analytics.

The stability of the free surface of the half-space with the mass flux at the boundary was studied.

Acknowledgements

This work was financially supported by Government of Russian Federation, Grant 074-U01.

Author details

Tamara Pogosian[1], Ivan Melikhov[1], Anastasia Shutova[1], Alexey Amosov[2] and Sergey Chivilikhin[1]

1 ITMO University, Saint-Petersburg, Russia

2 Corning Scientific Center, Saint-Petersburg, Russia

References

[1] Happel S.J., Brenner H. Low Reynolds Number Hydrodynamics. Englewood Cliffs: Prentice-Hall; 1965.

[2] Frenkel, J. Viscous flow of crystalline bodies under the action of surface tension. Journal of Physics 1945;9(5) 385-391.

[3] Tanveer S., Vasconcelos G.L. Bubble Breakup in two-dimensional Stokes flow. Physical Review Letters 1994;73(21) 2845-2848.

[4] Jeong J.-T., Moffatt H.K. Free-surface cusps associated with flow at low Reynolds Number. Journal of Fluid Mechanics 1992;241 1-22.

[5] Pozrikidis C. Numerical studies of singularity formation at free surfaces and fluid interfaces in two-dimensional Stokes flow. Journal of Fluid Mechanics 1997;331 145-167.

[6] Hopper R.W. Coalescence of two equal cylinders: exact results for creeping viscous plane flow driven by capillarity. Journal of the American Ceramic Society 1984;67(12) 262-264.

[7] Rodygina O.A., Chivilikhin S.A., Popov I.Yu., Gusarov V.V. Crystallite model for flow in nanotube caused by wall soliton. Nanosystems: physics, chemistry, mathematics 2014;5(3) 400–404.

[8] Lamb H. Hydrodynamics. Cambridge: University Press: 1932.

[9] Levich V.G. Physicochemical hydrodynamics. Englewood Cliffs: Prentice-Hall; 1962.

[10] Voeght F.D., Joos, P. Damping of a disturbance on a liquid surface. Journal of Colloid and Interface Science 1884;98(1) 2--32.

[11] Voinov O.V. Quasi-steadi-state flow in a liquid film in a gas: comparison of two methods of describing waves. Journal of Applied Mechanics and Technical Physics 2007;48(3) 385-392.

[12] Antonovskii L.K. Interface boundary dynamics under the action of capillary forces. Quasisteady-state plane-parallel motion. Journal of Applied Mechanics and Technical Physics 1988; 29(3) 396-399.

[13] Chivilikhin S.A. Plane capillary flow of a viscous fluid with multiply connected boundary in the Stokes approximation. Fluid Dynamics 1992;27(1) 88-92.

[14] Chivilikhin S.A. Relaxation of a small local perturbation of the surface of a viscous fluid in the Stokes approximation. Fluid Dynamics 1985;20(3) 450-454.